50 Ways to Fool Your User

From the way we interact with our workspaces to the simple act of changing a duvet cover, the world around us is shaped by design, and not always for the better. This book offers an engaging look at how everyday objects and systems can confuse, frustrate or even hinder us, yet also explores how a better understanding of human behavior can lead to improvements.

Written with humor and professional insight, *50 Ways to Fool Your User: How to Make Everyday Products and Systems Work for Us* invites readers to question the quirks of modern life while imagining how things could work better for everyone. Across 50 chapters, scientific explanations are paired with snappy anecdotes. Each chapter concludes with actionable takeaways. Whether it's struggling with unwieldy packaging, enduring the infamous middle seat on an airplane or navigating the frustrations of an AI call center, these relatable scenarios highlight the often-overlooked aspects of design that impact our daily lives. In the final chapter, the ideas are summarized into a neat practical ethos, offering ergonomic principles to inspire smarter, more thoughtful solutions in everything from technology to office furniture. Through reading this book, the reader will gather a view of what good and bad design looks like and how these examples can inform their work in designing better products, systems and services.

This book is for professionals and academics interested in human factors, ergonomics and designing with the human in mind, but it is also interesting for every layman. It will appeal to designers, engineers and systems operators.

Peter Vink is a Professor at TU Delft's Faculty of Industrial Design Engineering, the Netherlands, specializing in environmental design. From 2015 to 2021, he led the Sustainable Design Engineering department, overseeing 110 researchers and educators. He has written over 250 publications on comfort, performance and interior design. He is ranked as the second most-published author in Applied Ergonomics, the most important journal in his field, and he has guided over 28 PhD students and delivered 50+ global keynotes. He is a past President of Humanfactors.nl and received the Hal W. Hendrick Award for outstanding contributions to the human factors and ergonomics field.

Alan Hedge is an Emeritus Professor at Cornell University, Human Centered Design Department, USA. His work focuses on workplace ergonomics, addressing health, comfort, and productivity through research on workstation design, computer ergonomics, and environmental stressors like indoor air quality and lighting. He has authored or co-edited four books, *Advances in Ergonomics Modeling and Usability Evaluation*, *Ergonomic Workplace Design for Health, Wellness, and Productivity*, Keeping buildings healthy: How to monitor and prevent indoor environmental problems and *Handbook of Human Factors and Ergonomics Methods*. He has authored 41 chapters and over 270 journal articles and conference publications. He chaired the National Ergonomics Conference for 14 years. He is a US-Certified Professional Ergonomist and a UK Chartered Ergonomist. He is a Fellow of the Human Factors and Ergonomics Society, the International Ergonomics Association and the UK Chartered Institute of Ergonomics and Human Factors. Alan has received prestigious awards, including the Oliver K. Hansen Outreach Award and the Alexander J. Williams Jr. Award from the HFES for his impactful contributions to ergonomics and human-centered design.

50 Ways to Fool Your User

How to Make Everyday Products and Systems Work for Us

Peter Vink and Alan Hedge

CRC Press
Taylor & Francis Group
Boca Raton London New York

CRC Press is an imprint of the
Taylor & Francis Group, an **Informa** business

Designed cover image: Peter Vink and Alan Hedge

First edition published 2026
by CRC Press
2385 NW Executive Center Drive, Suite 320, Boca Raton FL 33431

and by CRC Press
4 Park Square, Milton Park, Abingdon, Oxon, OX14 4RN

CRC Press is an imprint of Taylor & Francis Group, LLC

ISBN: 978-1-041-06783-2 (hbk)
ISBN: 978-1-041-06641-5 (pbk)
ISBN: 978-1-003-63703-5 (ebk)

DOI: 10.1201/9781003637035

Typeset in Times
by Deanta Global Publishing Services, Chennai, India

Contents

Preface

Everything around us is designed and engineered. The car we are driving, the house we live in, the computer software we work with, the light in our room, and even the work we do. However, not all the things and systems we use in our daily lives are helping us. Some even hinder us in our daily activities. This book is written in a conversational style that inserts humor to help us recognize the silliness of many of our everyday experiences. Yes, I laughed out loud when recognizing some of the ridiculous situations that we put up with on a daily basis. Yet, the style is also informative, with a fair smattering of "hard" science for those who want to know more. Peter Vink and Alan Hedge take us on an insightful journey, unpacking bad design and engineering through a wide mixture of contemporary everyday personal and workplace experiences. But don't be fooled, the authors don't simply take the hard science at face value (even the hard science that the authors participated in). Expect the "hard" science to face the same critical (and humorous) eye as poor design and engineering.

Finally, the book doesn't just ridicule designers, engineers, and scientists for creating a world that is so frustrating. It also helps us understand how it could be better, with many apparently simple solutions that just seem like common sense. Once again, don't be fooled. Common sense obviously isn't that common otherwise we wouldn't be so frustrated with all the silliness around us all the time. The authors introduce readers to a profession whose job it is to understand how humans behave and how we experience things when interacting with the world around us in order to improve the "design". In short, the authors introduce readers to ergonomics and human factors (a term you have probably heard of but usually associate almost exclusively with a design slogan or selling point for a keyboard or a chair). This book is therefore also an attempt to give some support to those designing, engineering and making things and systems. It is written by two professors who have been busy for many years in the field of ergonomics and human factors (and design). They both have wonderful (and long) resumés working on the science and implementing and testing designs to see if they improve our lives. This book is a wonderful opportunity to learn from their lifelong experiences.

As the president of the International Ergonomics and Human Factors Association (IEA), I encourage you to read this book, even if you just want to have a laugh at the world (and yourself). You will recognize many everyday situations and might gain some understanding of designing with human factors and ergonomics in mind. This book is a "must-read" for everyone who experiences even a little frustration with the engineered world around us (which, let's face it, is everyone). At the very least, you will come away knowing that you are not the only person who is frustrated and that there is a whole army of people out there who have dedicated their lives to understanding how humans behave and experience things so that we can design a world that is a less frustrating place.

Prof. Andrew Thatcher
President, International Ergonomics and Human Factors Association (IEA)

1 The Best Working Place in the World

1.1 TOILET PAPER

When you aren't at home, on a busy day at work or traveling, it can be difficult to find the time to go to the toilet. But mother nature loves to play tricks, and when there is no time, you always have to go! So, about to enter a business meeting, you want to be comfortable, so you think "let's do it" and you go to the nearest bathroom while there's a little time for it, and if you do things quickly it should be doable. Relief is in sight. You open the toilet door and close it, of course. It would be strange to keep the door open in our Western culture, but not in all parts of the world. You loosen your clothing. Oh dear, today that's also more difficult than normal. So, it takes a bit longer. But you think, "well, just let the people in the meeting room wait". Finally, you're seated and you relieve yourself. Phew, that feels better, off to the meeting! But then the mystery and distress begins. Where is the toilet paper (again more commonplace in Western cultures)? Ah, you see something on the wall next to the toilet. It has a window so you can see it has toilet paper. But your first attempt to get a piece of toilet paper fails. In the device where the toilet paper is, the beginning of the roll cannot be found. It's an unclear roll anyway; apart from the narrow window slot, there's nothing to see. You have to try to find a piece of paper by touch. There is enough scientific evidence that visual information is important for user convenience (Norman 2013), but for that, you need to see where to grab the paper, and the designers of this device apparently have not yet discovered this knowledge! You can see that there is toilet paper but you can't see where the starting point of the toilet paper is. You reach under the device and try turning the entire roll a few more times to see if that could help. But that leads to nothing. You think "I'm in a hurry", but going into the meeting with sticky, smelly underwear is not ideal. So, let's try again. Oh dear, turning the roll around doesn't solve anything. You begin exploring your options. Maybe you can see where the starting point of the next toilet paper is if you lay on the ground with your head looking upward into the toilet paper device. However, the toilet space is small and there is not enough space for you to do this and also the toilet stall isn't well lit. Hmmm, maybe you can try to open the round box that contains the roll. But that doesn't work either – it looks like you need a special tool. Hitting the holder also doesn't produce any real results. And the people in the meeting now are waiting for you to arrive. What to do? Should you call facility management? No, that's probably a bit ridiculous. What would they think of you? And then they would have entered my smelly cubicle. Anyway, you don't see their number anywhere. No, that's not going to work. Maybe it is a user error. Maybe if you put your hand in the device from the bottom again and then move your fingers

DOI: 10.1201/9781003637035-1

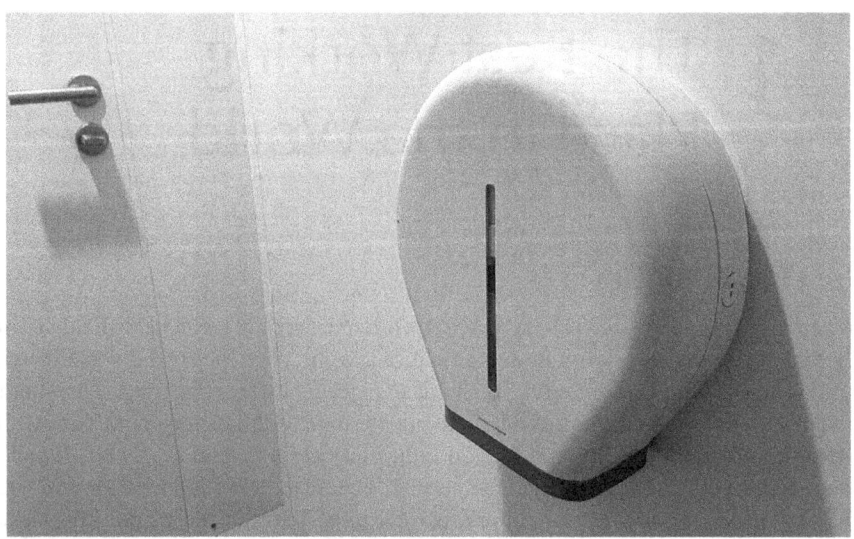

FIGURE 1.1

across the roll of paper, you will find the starting point of the paper roll. Again, no success. Wait, if you scribble on the toilet paper maybe it will release the outer layer of paper and perhaps then you can pull the paper loose. After a lot of fiddling, you remain confident that it will work. Alas, doing it frees not just one layer but several, which disrupts the entire system and leaves scraps of paper falling on the floor, but at least now you have a piece of paper to wipe off, and you can still fetch another one from the broken roll. Fortunately, you try to pick up all the paper from the floor and push some of the paper back into the holder, but that is not successful. Just throwing the scrap paper away is the only solution. Hopefully, simply flushing the paper scraps will still work, but not in every country, like in Greece, because paper can cause a clog in the drain. But there is no obvious waste bin, so maybe you should leave the paper scraps strewn on the floor? And now you also just have to come up with a story about why you are late for the meeting. Of course, it sounds silly if you tell attendees that you don't know how this toilet roll works, and displaying your "ignorance" is not conducive to your position as an ergonomic designer in the enterprise. The toilet roll holder clearly has been designed to be visually "elegant", to be a smooth shape for easy cleaning, to have an inspection window, to see whether a new roll is needed and to be securely locked to prevent people from stealing the toilet paper roll. But it is doubtful that any user testing was ever conducted by the designers of this product.

In ergonomics, there are many examples of the importance of visual information in performing a task. For instance, Vignais et al. (2013) even found that direct visual feedback on the physical posture of employees ensures that risky postures are reduced. However, not everyone can see well. For a visually impaired or blind user, the toilet roll holder design fails to provide any tactile cues on where the toilet paper is dispensed and/or how much is dispensed. As described by Norman (2013) the designers should at least have conducted some user testing to check how well it works in several situations e.g., different user groups (men, women, children, different cultures, etc.).

1.2   FIRST CREATE A CALL

Today you are working from home. You finally have arranged all signatures within your organization, and at work, you had created a beautifully designed quote that is now ready for the customer. You are proud of your work. It has to be at the customer's at 12:00 today. It's 11:00 am and you're working from home so you have plenty of time for this. Just log in and then the important quote can be sent to the customer. But what is this, when you try to log in there is no connection with the office. The misery starts. Well, we have a helpdesk. So, they will solve the problem. You just give a call to the helpdesk. You will explain the situation and then you expect that help will be given, but to your surprise when you call the helpdesk the answer is:

Helpdesk: "You must first create a call".
You: "But I have you on the phone now, can't you help me now?"
Helpdesk: "No, that's not how our helpdesk works, you first have to create a call, which you do by sending an email to the helpdesk".
Me: "But I can't do that, because I don't have a connection from home".
Helpdesk: "I can't do anything without a call, does your neighbor have working internet? Or can you send an email from your phone"?

So, now you have to send an email to the helpdesk! How helpful is this helpdesk? Has madness struck the IT world? The time people spend making the call, creating

FIGURE 1.2

the system in which this call fits, evaluating the call to gauge satisfaction and being able to check what people have done are all more important than helping colleagues within the same company and who work in the primary process. Working from home is all fine, as long as you can electronically access your files over a network. But if there is an IT malfunction at work or there is no connection for some other reason, a well-functioning IT department that can help quickly is a requirement.

Fortunately, this time you were able to send something via your mobile and private email. Then you had to call again to check whether they had received your email, but this time, you got another "help" employee. So, now you had to tell them the whole story again and end it with "I have created a call, can you help me now?"

Helpdesk:	"Just wait, it's not in the system yet".
You:	"But I would like to finish a job quickly and I need to email a quote to a prospective client".
Helpdesk:	"We are all busy of course, please be patient. As soon as the call is received I will call you or else my colleague will call you".

It's now 11:40 on your phone so you try again. Yep, you get another "help" employee, so you explain the whole story again. The helpdesk answered: "I'll find out". Now time is passing, and you feel that things are getting out of hand. So you try another solution. You just call the department secretary. She understands you and she sends the quote!

The helpdesk system for creating the call costs a lot of time, frustration and annoyance. Whoever designed the call system must have had good intentions, but in this case, simply calling a colleague and getting immediate advice was the fastest solution.

Sarcastically, it seems to you that the fire brigade and ambulance should also adopt this system design. Just driving up to a fire and putting it out would just be ridiculous. No, let the owners of the burning house create a call first by email and talk to various persons who work at the fire brigade and then they will receive help.

Keller et al. (2020) confirmed the problem that having to wait at work is an important stressor. They state: "Work interruptions come in various forms, such as emails, equipment malfunctions, or colleagues seeking a listener". In recent large-scale studies, employees reported work interruptions as the most frequent work stressor (Baethge et al., 2015). Wajcman and Rose (2011) found that, on average, knowledge workers were confronted with 85 work interruptions during their workday. With the rise of new technologies (e.g. smartphones), the prevalence of interruptions seems to have increased even further, becoming more and more relevant to contemporary working life. We have to go back to making systems work and helping each other instead of designing systems that make work more complex. The whole idea comes from someone, and it is implemented in many enterprises. Maybe it is time to rethink these designs.

1.3 WORKING FROM A DISTANCE

Have you ever tried to work with a laptop in an economy class airline seat? It gets especially interesting when the person in front of you reclines their seat. The screen of the laptop is now bent toward you so that only the upper edge is visible. By tilting your head downward and a bit sidewards in a precarious position, you can see a bit of the text on the screen. In addition, your wrists now must be bent 90 degrees, your elbows cannot be to the side, they have to be in front of your stomach because other people are sitting next to you. So, all joints of your arms, your head and neck, are in awkward postures. In short, discomfort, aches and pains can arise very quickly in this position. It is very clear that there is discomfort in all kinds of places in your body. The question is how long you can keep this up? Working like this is simply not possible in economy class on an airplane without increasing the risk of an injury. That certainly applies to people who are a bit more rotund and/or are taller.

But business travelers need to work, and paying for a more spacious business or first-class seat often is not an option. Those who can and want to work while traveling need help. Working in the office is often still possible because most workplaces usually meet ergonomic standards. Although that is not always the case, let's assume it is, OK. Let's assume that the information and communication technology (ICT) systems work. Oh wait, this is a dangerous assumption. Because if the ICT systems work, the question is whether they do exactly what is needed for the task. The malfunctions, the start-up, the unwanted settings and all those login codes alone have

FIGURE 1.3

nothing to do with the content of the work. It should be possible to make this easier, you would think. But well, while these may work better in the office this may not be the case while you are on the road. In your office workstation the height of the desk, the chair position, the lighting, the noise and the air quality can in theory all be arranged to be quite good. Let's also assume that there are meeting rooms and concentration rooms when they are needed for the tasks in the office. However, working on the road is more difficult. Although reliable and fast Wi-Fi is often available when you travel, so you can work well on a train or an airplane or in a hotel, such connection is not yet universal. But even at more expensive hotels, a fast wireless connection sometimes is not included in the price, or the available systems are slow. It's also extremely difficult to work on buses, especially on bumpy roads. Trains often are too crowded to work during rush hour. There is noise, people bumping into you, but even on transportation that isn't full, your neighbor usually gets the idea to start talking on the phone and speaks very loudly. It's often a pointless chatter. "I think I am on my way to you now. I'll be home by six, if everything goes according to plan". It's nonsense. It can be sent by text, no need to make a phone call and talk so loudly, but you still cannot avoid listening to it and it's annoying. Simply reading on a bus or train is not always easy and often you are surrounded by sleepers, or the vehicle is so full that there is no place to sit at all. According to a study by Robertson and Vink (2012) conducted pre-COVID, 20% of Americans work from home and most are not satisfied about their workplace at home. The same study shows that people work in the kitchen, on the couch and (luckily) also on a desk. There are observations where people place the laptop on the ironing board. There are laptop stands available and separate keyboards improving the neck position, but these are certainly not always used. The home office often doesn't meet ergonomic standards: no adjustable tabletop and chair, screen not in the right position, etc. Internet connections often don't work either, because the provider is messing around, bad lighting, no good storage facilities and it's difficult to access work files on the net. We also need to stand up for those who are willing and able to work. A clean task for facility managers, product developers, organizational experts, software developers and management. All these issues were greatly magnified by the COVID pandemic where working from home was mandatory for many. Bureau of Labor Statistics (2023): in 2022, some 27.5% of the US workforce was working remotely at least part-time (US BLS, 2023), while some surveys estimate this to be as high as 50% (Walsh, 2023). Today, many companies are experimenting with only partial returns to the office, maybe for three days a week, and this "hybrid" working is becoming more popular with 41% preferring this way of working (Parker, 2023). While technology has become more portable, networks have become more reliable and most offices have ergonomic designs, for many people the challenges of poorly designed settings for remote work remain.

1.4   COLOR, PLANTS, LIGHT AND PRODUCTIVITY

Look, we all want to know exactly how our environment increases productivity. Just imagine. We adapt our environment a little and we only must work at half strength for the same results. Isn't that great? Maybe you can now finish cleaning your room

FIGURE 1.4

twice as fast with good music playing. Maybe you can now complete an assignment in half the time with the right color of your room. Your boss sees the final report a week earlier than expected because of the space design with the right shapes, and in a bright yellow garage, your car is repaired in half the time. In 2014, Iris Bakker obtained her PhD at the TU-Delft. It was about productivity and the environment, and she tested the effect of working in a red and blue meeting room. The press was excited to hear great effects. The title of her dissertation is: "Uncovering the Secrets of a Productive Work Environment: A Journey Through the Impact of Plants and Colour". However, the PhD showed that quite a lot of nonsense is written. Often the effect of environments is exaggerated.

Most people like flowers and greenery. There is some evidence that word association performance improves if there are plants in the space (Hesselink et al., 2004). However, Larsen et al. (1998) found that in an office randomly altered to include no plants, a moderate number of plants and a high number of plants, self-reported perceptions of performance increased relative to the number of plants in the office and higher levels of mood, perceived office attractiveness and some perceived comfort improved when plants were present, but performance on timed decision-making and cognitive processing productivity tasks actually decreased with the number of plants in the office.

Although it is often assumed that plants can improve productivity by improving indoor air quality, a critical review of results of both laboratory chamber studies and field studies concluded that indoor plants have little, if any, benefit for removing indoor air of VOC in residential and commercial places (Girman et al., 2009), Hedge, 2016b). If you like plants then in general they might have a small positive effect on your mood.

However, other work has revealed some very interesting findings about being immersed in nature. Park et al. (2007) showed that walking in the forest compared to walking in the city has some beneficial effects on the human body. The cerebral activity in the prefrontal area of the forest area group was significantly lower than that of the group in the city area after walking; the concentration of salivary cortisol in the forest area group was significantly lower than that of the group in the city area. Later they checked 32 scientific papers with randomized controlled trials; health outcomes were collected from 6264 participants aged 6–98 years. The interventions in these studies were walking (n = 21), staying (n = 7), exercise (n = 4), indirect exposure (n = 4) in the forest and the activity time was between 10 and 240 minutes. Overall, walking showed consistent positive health effects.

The effect of color is also unclear. Also, because there are so many colors and shades of color and, depending on the background and color source, whether the color is reflected or luminous, we also see it differently. A study by Bakker (2014) on the effect of the color of the meeting room was disappointing. It was a complex study studying the effects of 250 meetings in three different meeting rooms (red, blue and control). A completely red room or a completely blue room had no significant effects on the meeting quality or outcome. The effect of the facilitator was much larger. Similarly, Chung (2015) performed a series of experiments using different illuminance levels, task luminance and color temperature and found no effects on creative performance.

However, a study of 1,077 people showed that white is a preferred color for the interiors of buildings (Bakker, 2014). The same study indicates that white is also pleasant for concentration and there are some studies that show that red might make people feel more energetic. For meeting rooms, white is indicated as the most preferred color by 31%, but many people have no preference (23%). White is also the most preferred color for the office, but green and blue tones are also mentioned (total of approximately 20%).

Other research by Hedge and Sakr (2005) has already indicated that ergonomically correct furniture can increase productivity by 4.5%. Of course, it depends on the starting situation. In a bad old situation, the improvement might have a large effect. If the starting situation is already good, that 4.5% might not be achieved.

So the design of the environment alone doesn't always determine one's mood or performance, but attention to ergonomics and environmental design remains important, and it is beneficial to add a lunch break while walking in a forest or park and provide decent light.

1.5 CONTROL

"Isn't it great"? Imagine that you are a facility manager, and you can predict energy consumption in your building. The light goes on and off automatically and the temperature and ventilation are centrally controlled. The white noise can also be arranged. This might mask the noises and sounds from elsewhere in the building. You have control over the building. Every room and every corridor has a centrally organized environment control. Everything can be arranged from one place in the building. You can make it dark when no one is in their room. This can be detected by cameras, or even automatically, when there is no movement in the room, the light dims and the heating or cooling stops. In this way, much energy is saved. Having so much control and power is amazing. It is a pity that there is still a group wandering around in the building who also want to have control over the indoor climate: the

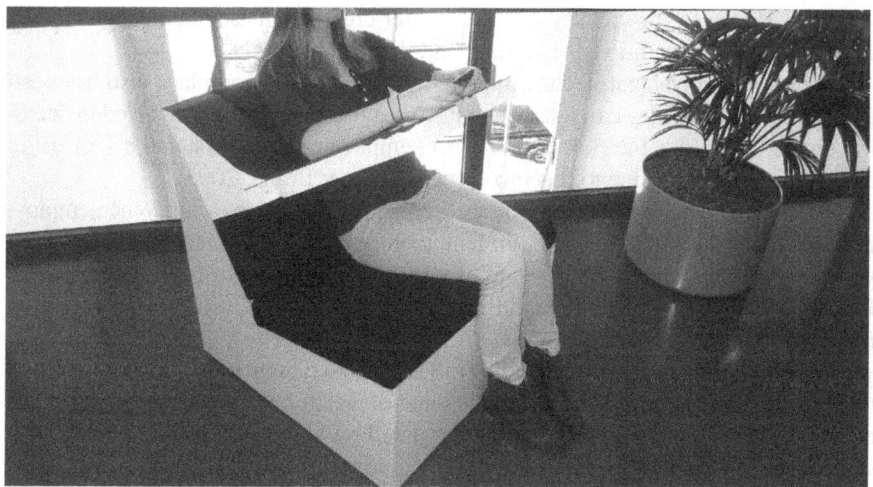

FIGURE 1.5

employees. And there are many more of them than the number of facility managers. They also contribute to earning the money or producing the output. So they are quite important. But the indoor environment of buildings is a complex constellation of environmental factors, only some of which could be directly controlled by employees (O'Reilly et al., 1998), especially in large, tall buildings where operable windows are impractical. Hedge (1993) tested the benefits of office furniture that incorporated a high-efficiency air cleaning system and showed that it improved reports of comfort and health, but most buildings don't do this. In the mid-1990s, the US company Johnson Controls developed the "Personal Environments" module (PEM) system where each user has fingertip control of air temperature, air supply, radiant heat, background noise and task lighting at their desk with a unit connected to the building air-conditioning system. When the workstation or office is occupied, power is provided to all functions. When the area is unoccupied for approximately 15 minutes, as determined by a sensor located in the "User Control" unit, the power to PEM shuts off and it enters standby mode until the User Control Unit sensor again detects an occupant (Bauman & Arens, 1996). In spite of much research showing the benefits of the PEM approach, it never gained commercial traction. More recently, Kong et al. (2019) showed that a stand-alone microenvironmental air conditioning unit can be placed at each desk to give employees control over their thermal and air quality conditions. Whether any companies will use this kind of technology remains to be seen.

Bazley's dissertation (2015) shows, like many other studies, that while it is essential for these employees to have control over the indoor climate, few actually do have control. Many of Bazley's studies have shown the beneficial effects when employees simply have the option to open a window; indeed, Hedge (1982) found that 77% of workers in open-plan offices would like to be able to do this so that can control the light themselves and also control indoor air quality if it is too hot and stuffy or "smelly". However, for security reasons and for creating environmental consistency of indoor climate conditions, few office buildings allow this.

Employee involvement in changing the interior of their workplace also has positive effects because they can test different workstations and can be involved in deciding on the new office interior. Performance, employee satisfaction and even trust in management improve due to this involvement (Lee & Brand, 2005).

Improvements in productivity and well-being also can also occur. When employees are exposed to more variation in temperature, they feel a greater range of temperatures as comfortable. Giving employees the feeling of control therefore pays off. Dear and Brager (2002) also found this. They compared employees who worked in naturally ventilated buildings, where windows can be opened, with employees who worked in centrally controlled buildings. People who can open windows get used to the variation in temperature, and their range of what they experience as comfortable is greater. Employees in naturally ventilated buildings also tend to report fewer symptoms of "sick building syndrome" compared with those in air-conditioned buildings (Hedge et al., 1989).

But if employees have individual control of their environments, then as a facility manager you now have a slightly tougher time, because you must give up some control and look at how you can increase the sense of control for the employees without compromising any laws and standards that govern what environmental conditions should be provided to employees. And there is more "input" from the end user in the adjustment or purchase of environmental systems.

There are also consequences for the spatial design. Honan (2015) investigated how employees in a pharmaceutical company used information and communication technology and where they preferred to do this. The smartphone and tablet were used together for about two hours per day, and employees preferred to use these on a "couch or non-office chair", a design often unavailable to office workers. For the workforce studied this means that more "lounge-like" spaces are needed in those buildings. Nowadays there are even special tablet or smartphone chairs. Veen et al. (2012) found a significant reduction in neck flexion when using a special seat where the smartphone or tablet is positioned higher. This is an important intervention as neck pain can be a problem using smartphones. The incidence of neck pain among smartphone users varies widely, with studies reporting prevalence rates ranging from 17.3% to 67.8% (Mathew & Walarine, 2020). Yalcinkaya et al. (2020) found that students use smartphone between 3.6 and 4.2 hours per day, and there was a relationship between the daily calling time on the smartphone and neck pain. When there are special areas for using the smartphone or tablet and if employees can go to different workplaces in the building, then they have control over the environment and can choose to go there. They can then work in an armchair with their smartphone or tablet. That too becomes an additional task for facility managers. To find out how the different changing tasks can be supported by the environment without increasing health risks for employees they must find out what tasks are done, what the ideal environments are for these tasks and calculate how many of these spaces are needed ideally. Maybe even a planning system is needed to facilitate employees knowing when which space is available. This means facility management is becoming a more complex, but also a more interesting job, including giving control to employees. And the challenge for Ergonomics is to develop products and conditions that provide employees with a safe and comfortable workplace, wherever they are.

1.6 MEETING MISERY

According to legend, the English King Arthur sat his knights at a round table – why? In a study of different meeting tables, the oval or round tables performed better than the square or rectangular ones. There were significantly more interactions and the quality of the outcome of the meeting was better using the round or oval meeting tables (Kooij-de Bode et al., 2009). The meeting atmosphere was the least good at the square or rectangular table and now comes the interesting part: we know it. One hundred fourteen people were shown pictures of four meeting setups: square/rectangular, round/oval, a kind of bar stool setup (high table and high chair) and a sofa (homely atmosphere). Almost everyone indicated that the rectangular/square is less good for meetings. For example, the 114 participants mentioned that the cohesion in the group is by far the lowest at the rectangular/square meeting table. But it's not just table shape that is important, it's also where you sit for the task you are doing. If you want to converse or cooperate or compete with a colleague then at a square table you sit opposite them or corner-to-corner facing them but at a round table you sit next to them to cooperate or converse and opposite them to compete (Sommer, 1969). To minimize some of these effects at the UN the table is a U shape.

Yet for space efficiency, we usually meet and/or eat at a square or rectangular table. So the misery is that we mostly meet in rooms with square or rectangular tables. Apparently, designers must think that we like to meet at an unpleasant table. Maybe we seek unpleasant situations. Maybe we look for misery, for something to complain about.

Or maybe we just don't care about always having a pleasant atmosphere or good meeting results. We might be looking for an argument! And it's not just the table

FIGURE 1.6

shape that's important, but also features of the meeting room that can have an effect on productivity, creativity and communication. Six groups (teams) had several meetings in the four different setups (in total 120 meetings). After each meeting, participants completed a questionnaire. The results were the same as those for the 114 persons reported based on just seeing the meeting room (Kooij-de Bode et al., 2009). The meeting atmosphere was significantly lower (average 4.4 on a scale from 1–7) in the square/rectangular condition than in the others (5.4 in round/oval, 5.3 in bar stool, 5.1 in the sofa setup). Also, group cohesion was significantly lower (average 3.9 on a scale from 1–7) in the square/rectangular condition than in the others (4.9 in round/oval, 5.2 in bar stool, 4.8 in the sofa setup. Seeing each other (needing to see each other's facial expressions) was also lowest at the rectangular meeting table. In fact, we know it, but still, the majority of meetings are at rectangular meeting tables. Probably, we are used to it, or it is easy to enlarge the meeting table or make it smaller with rectangular tables, or the table occupies less space. Or maybe we deliberately seek misery.

Other effects of interior design have been established as well. For instance, Dazkir (2009) found that more positive emotions were stimulated by meeting tables with sloping ('curvilinear') shapes than with straight lines. In the round and oval shapes, viewers were also significantly "happier" and "relaxed" than in other shapes. Xylakis et al. (2021) showed that properties of the form such as curved or complex spaces align highly with increased arousal. An earlier study (Kinch, 1982) had showed that placing furniture closer together increases the number of social interactions, up to a limit, especially if you don't like the adjacent person! Many spacing guidelines are available for workspaces, conference rooms etc., as a quick web search shows you. When heads are more than 2.5 m (8 feet) apart, the number of interactions decreases significantly. We should be careful with generalizing this result as the study focused on the social interaction of institutionalized elderly. However, for meetings, it may not be wise to place chairs too far apart, but also you don't want them too close like in airplane seats. In lounge environments, where communication needs to be promoted, it is also advisable not to place chairs too far apart. If the chairs are about 1 m wide, maybe there should be a maximum of 1.5 m between them. Incidentally, chairs with their backs to each other also result in significantly fewer interactions, which, of course, is not really a surprise. When two people communicate an angle of 90 and 45 degrees between them seems preferable rather than sitting opposite to each other or parallel to each other (Piro et al., 2019). So, for many kinds of meetings, including dining, the best advice is to look for the possibility to have a round or oval table where participants can see each other's facial expressions.

1.7   MANAGERS ARE PAID TO SUPPORT

Enthusiastic employees perform better work (Bakker, 2014). Bakker also bases this on management guru statements. Some caution is required here, because those gurus often base their statements simply on a belief and that is the opposite of science and using big data to find out what is true. With beliefs, there are no data used, and it is hard to track the source of the belief. Sometimes researchers analyze case studies

FIGURE 1.7

afterward and confirm their belief with that subjective information. But it also seems reasonable to think that enthusiastic employees perform better for certain types of tasks. But is it really universally true?

In 2015, Gallup published a report based on a study in more than 150 countries among 2.5 million teams led by managers. Well, you might think, that's a lot of data, but with the internet, a large group can be gathered rather quickly these days. So, should we really be influenced by the big numbers that may be self-selected and may not be a true representative sample. This is not about someone who conducted 2.5 million two-hour in-depth interviews on their own with a statistically valid sample frame. That would have taken 2,400 years with a 40-hour workweek, and as far as we know, even though some sharks can live to be 400 years old, people can't live 2,410 years! The added ten years is because we assume that conducting in-depth interviews about management styles won't be conducted on a sample in the first ten years of a person's life. By the way, travel time is also ignored. Suppose we do not ignore it, and all countries had been personally visited by the researchers so they could sniff the culture and interpret the data better, then the time required for the survey would have been even more. Even with internet data (big data) and using AI, it could have been a lot of work, and while we can draw some conclusions, the depth of the data will be limited, and some caution is required when interpreting results. Of course, there still are interesting findings in this report.

This Gallup report shows how to increase the chance of having passionate employees: 67% of employees became passionate when their managers focused on their strengths, but only 31% felt passionate when managers focused on their weaknesses. So, it is better to focus on strengths. Reinforcing strengths is better than punishing weaknesses. That seems obvious. However, some employees also work well with different encouragement. We are quickly inclined to see what is not going well and indicate to colleagues what can be improved. But it is, in general, better to focus on what is going well and to expand on that. This also applies to managers, who must focus on the strengths of employees and expand on that. If you have a team of people who all have strength in what is needed for the results of the department and you know how to use that as a manager, you are more of a leader, according to that report. But it does happen that managers are appreciated when they can show what the performance of the people is and can indicate what is not going well and what they encourage their employees to work on. Something like: "Jack, you are very bad at giving a presentation to a room of people. You have to get better at that". That doesn't seem to work, but the real leader says:

> Hey Jack, you did a really good job with that analysis, shall we look for opportunities where you can do more analyses and present that in an attractive way as well or maybe when you hate presenting check if we know someone who likes presenting?

The real leader can use employees' strengths for the whole and achieve results with that.

A lot has changed since the 2015 report was written. COVID has changed workers and workplaces. Hybrid working has become more popular. Statista (2024) reported that in the second quarter of 2024, 53% of US workers are working hybrid, sometime in the office and sometime at home. When working remotely, it becomes more difficult to see the strengths of employees. Only looking at the results based on (big) data is tempting for a manager and it is fast, you don't have to travel, you sit behind your PC and see the data and you can show the data to your manager. But a personal conversation in which the qualities are discovered and used as a whole is perhaps better for the company and those involved. Whether this can be demonstrated in a pure case-control study is very questionable. So for the time being, we can continue to use some big data studies and build on the strengths of our colleagues.

1.8   A PLACE OF MY OWN WITH SMART ENVIRONMENTS

Wow, that sounds cool! Your company is implementing a "New Way of Working". You will go to a new building. During the move, you will put your plant, statue and a few handy reference books in the moving box and then it's fun to move to a nice new environment. You are excited to go, but no one has given you a moving box. Whoa, whoa, that is not the intention, you've been told that the "New Way of Working" will be fun, but surely, you are not going to sit anonymously in a place where everyone else, even novice knuckleheads, also sit. Now everyone will get the same docking station for their laptop, the same screen, desk and chair. Even someone with a still-wet diploma gets that too! And, even more important, there is nothing of yourself around you. Anonymous and uniform. "Your personal workplace" no longer exists. Just a sea of bland uniformity. What a misery. Is this typically just Dutch? Research shows that it is not a local Dutch problem. A comparative study between the Netherlands and the USA showed that the same thing has happened in both countries (Robertson & Vink, 2012). Working from home was happening before the COVID-19 crisis. Twenty percent of the employees did it in the USA and 15% in the Netherlands. They also seldom had a good ergonomic workstation at home and there was no new style of management to deal with distributed workers. Now, in the USA and the Netherlands in the post-COVID return to work, offices are smaller with more

FIGURE 1.8

"shared desks" and more people are also working at home for part of their week. Indeed, it can be assumed that after COVID-19, many more companies in the world have adopted this "hybrid working" approach. In both the USA and the Netherlands pre-COVID, employees had a need for personalization of the workplace, which is also probably a universal need. But in many post-COVID workplaces, where workers may be in the office for a few days each week, there are shared "hot" desks, with limited to no personalization allowed. These days much "personalization" is simply changing pictures on the screen background. Perhaps there is a market for the manufacturer who can produce something that allows employees to quickly personalize their whole workstation. There should be an ability to personalize "quickly", because if you spend half an hour every day reorganizing your place, it will not be acceptable. Ideally, the workplace should recognize an employee and adapt automatically. That is possible, because, for example, some cars recognize the clothing colors of the person who gets in and automatically adjust the interior lighting (Wagner et al., 2014). Then it must also be possible to give the workplace personal characteristics. There are apps in the iPhone that ensure that your settings in the car (seat settings, dashboard layout, etc.) are adjusted to your personal dimensions and preferences. Then a lot must be possible in the office because your laptop or tablet contains much more information and with projections, moving elements, digital photo frames, ambient lighting and colors, infinitely more can be done than what is being done in the current situation.

Of course, there are possible solutions to create more of a feeling that it is your desk you are sitting at. Light, temperature, color and table height can adapt automatically when you and/or your laptop is recognized. Tackenberg et al. (2012) have developed a desk with a back wall, which separates two desks on which a projection of a photo, drawing, video or something similar can be shown, which is linked to the laptop (see the picture at the start of this chapter). When someone places their laptop on the desktop, this projection appears on the back wall; a nice example of the smart interior, where the workplace becomes more personal. Windows 11 can recognize your face and automatically adjust screen settings so it should be possible to also automatically configure features like the height of an electronic desk or the settings of a chair. It is not like your "dumb" old desk, so a personalized space should be possible with smart environments.

But for a majority of employees, it is still distressing to no longer have their own personalized workstation at work. There is clearly a need for making the work environment more personal and responsive to individual differences and preferences, and the future of work should be personalized (Tsipursky, 2024).

1.9   THE STATIC CHAIR

You can sit on almost anything, but don't ever work a day or more on a chair that is not adjustable, not comfortable and does not move. That leads to misery. In 2012, the results of an impressive office chair study were published (Groenesteijn et al., 2012; Ellegast et al., 2012). Many measurements were taken while office workers were sitting on five different office seats and doing prescribed office tasks. EMG, body posture, body movement, chair movement, comfort, emotional experience, personal preference, assessment of chair components and a subjective final assessment were some of the measurements. Ten participants sat on each chair for two hours and performed prescribed activities such as correcting text, filing, making phone calls, etc. In addition, 40 employees tested the five chairs during work for five weeks. The chairs were also selected in such a way that each person had a different chair order to prevent order effects. A great organization to perform these field measurements, but doing the tests did work after some pilot study.

And now the interesting thing. This entire 1.5-year expensive study led to nothing. It was a kind of occupational therapy for highly educated people. A group of highly developed people were just doing things, which were useless and had no impact at all. It also did not gain new knowledge as much of what was found was also described before. What nonsense you would say. We dare say this because one of the authors (the brilliant one) participated in it. At least that is the first impression if you look at the results superficially. Of the five chairs, four had a special dynamic property and were more expensive. But that did not matter because the control chair that had a simple synchro mechanism did perform also rather well. The added value of the newly developed chairs could not be observed in the many measurements. So, a standard chair is sufficient for many situations. Incidentally, interesting findings were made about all the different chair components, which were discussed with the makers of the chairs. Another result is that one chair was experienced as more comfortable and healthier by the test subjects and some body parts appear to move more in this chair. This chair has a seat that can be moved in all directions with damping, so that support is experienced despite the mobility. This chair therefore provides more mobility. A group of experts believe that more movement is healthier and a chair that stimulates this is preferred (e.g. Grooten et al., 2017).

However, Hedge and Ruder (2003) studied chairs with a dynamic chair backrest supposed to facilitate body movements while typing on a computer and found

FIGURE 1.9

no significant differences between chair back conditions (locked or unlocked for dynamic movement) in either the total number of movements or specific body movements over the duration of the test. They concluded that typing is a task that inherently requires a relatively static posture, and sitting in a chair with a dynamic back may not necessarily encourage greater movement. Use of the free-moving dynamic chair back did provide better back support for subjects when they changed from upright to reclined postures. Similarly, Hedge and Lawler (2007) compared the effects of one-hour sessions sitting on both a static seat (SS) and a rotary dynamic seat (DS). Half of the participants (total n = 36) had reported back pain at the start of the study. Results showed that while torso movement was greater for the DS condition, ratings of postural stability writing and typing, feelings of nausea and dizziness were worse for the DS condition and back pain was not reduced.

Interestingly, airplane seats that let you move have a positive effect. Bouwens et al. (2018) compared an interactive airplane seat with a current economy class seat.

In order to register the participants' in-seat movements, the bottom cushion of an aircraft seat was equipped with fabric pressure sensors. These sensors were placed between the upholstery and the foam of the cushion and connected to a laptop. By lifting the legs, participants could control a video game. Participants used both seats for 3.5 hours and performed significantly more in-seat movements when using the interactive seating system. Furthermore, this interactive seat predominantly led to significantly better comfort experiences. Passengers indicated that they would prefer this interactive seat over a standard aircraft seat. So, the effects of dynamic seating seems to be dependent on how much movement is required to perform the task and on the duration of sitting. In some situations, like writing, typing, dining etc., where body movement can be counterproductive a static chair can work well, while in other situations, where some body movement can be beneficial, a dynamic chair can be a better option.

1.10   OFFICE OR GYM

Soon you might no longer see the difference between a gym and an office interior. After a day of office work, you are physically exhausted because of cycling or walk-ing on a treadmill at your desk while reading documents, and also walking during meetings. To counteract the adverse health effects of too much sitting and increased activity, calory burns and brain bloodflow, which can also stimulate creativity, new ways of "active" works have been developed. You might come up with your best ideas during a thrilling activity like BASE jumping where, for example, you jump from a high building with a parachute to glide to your landing. The adrenal glands produce a lot of adrenaline, and if you are not frozen by fear new ideas can spurt out. Of course, if your parachute fails to open only blood spurts out! The fact that you land quickly ensures that you must choose your best idea quickly. The big soup of nonsense ideas is thus thickened into a broth drop with a brilliant idea. Of course, this is also possible in much safer situations without needing a parachute, and with-out the adrenaline surge but with the companionship of others because if your great idea is not captured or conveyed to others then that idea may quickly evaporate. Of course, these days if you have your phone with you, you can easily record your great idea for retrieval at a later time.

Unfortunately, organizations have hierarchies and when a manager asks employ-ees for their ideas, often these are not shared out loud because the employee fears criticism or ridicule for their idea if it is not in keeping with how management thinks things should be done. But new ideas are the seed of innovation and in today's fast-paced world, innovation is becoming more important to achieve good business results. So how can idea generation by employees be fostered? There are many books written on this topic but one theme receiving much attention these days is the impor-tance of increasing exercise at work to have healthy and vital employees. Research by Voss et al. in 2010, studied a group of "professional couch potatoes", has found that even moderate exercise (walking at one's own pace for 40 minutes three times a week) enhances the connectivity of important brain circuits, combat declines in brain function associated with aging and increase performance on cognitive tasks.

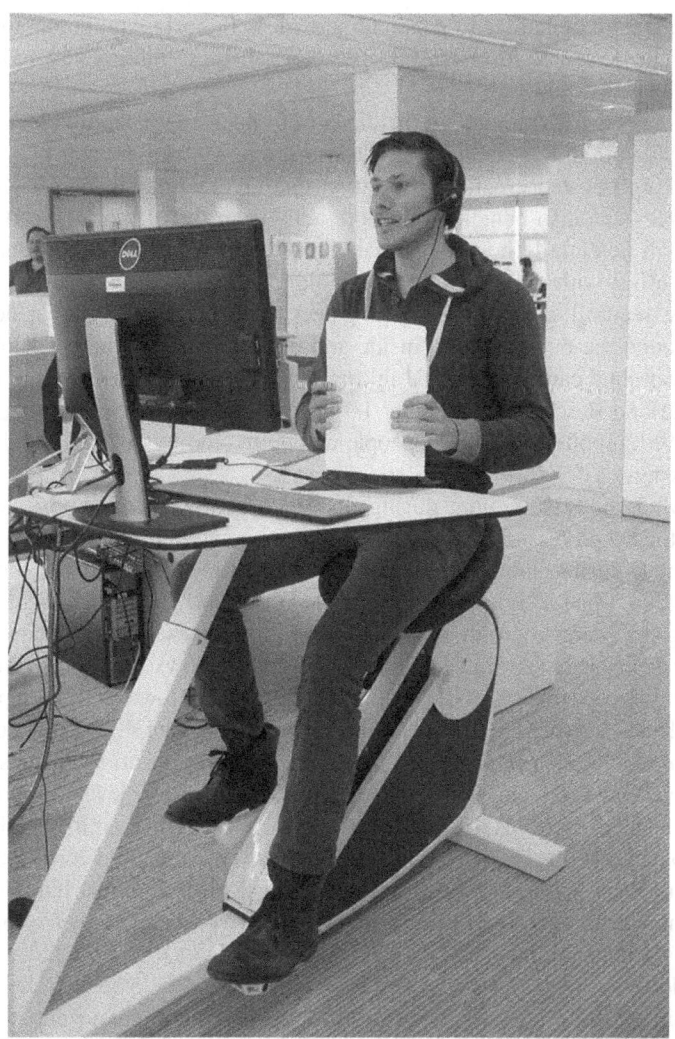

FIGURE 1.10

Walking 40 minutes at a leisurely pace three times a week can already improve the connections between important parts of the brain.

The process of idea generation and techniques to promote creative thinking has been extensively reviewed (Hedge & Lawson, 1979) found that while there are many techniques available to promote idea generation, there is only limited general awareness of these methods, and when others hear of great ideas, these often are ridiculed and those who have the ideas are labeled as "square pegs in round holes"! Also, people vary in their ability to generate novel ideas, some are naturally more creative than others, and the physical and social settings can either facilitate or inhibit idea generation. It may be harder to generate new ideas in boring, monotonous settings so varying the design of spaces, products, color schemes, lighting and activity may increase idea generation.

Enough has been written about the danger of under-exercising and measures are urgently needed in this area. Oppezzo (2014) has shown that for a majority of people tested walking boosted creative inspiration and creative output by 60% or more. So, employees can also benefit from more activity in the office. If an office has a gym that will potentially boost the creativity of employees who use it, perhaps the traditional office with only seats will eventually disappear and static desks will be replaced by treadmill desks and/or cycling desks, and employees will be encouraged to walk around the building both inside and outside in good weather. Going for a lunch walk around can boost your thinking, especially if you are walking in nature like in a park (as described in Chapter 1.4). Meetings can also be done partly while walking. With mobile technology, people can more readily engage in mobile work while walking.

Also, whole-body stretching during the lunch break can have positive effects. Sixty healthcare professionals performed a range of stretching exercises targeting the entire body during a lunch break period for three times a week for six weeks. The result was that it had a significant beneficial impact on pain and physical exertion. Comparing the exercise group with a control group the exercising group showed a significant decrease in pain intensity (mean difference 3.6 vs. 2.5) and physical exertion (mean difference 5.6 vs. 4.0) (Alqhtani et al., 2023). However, it is important to match the level of activity to the type of task being performed. Botter and colleagues (2013) compared task performance for six different workstation designs (sitting; standing; walking at 0.6 or 2.6 kph and an elliptical at level 4 or level 12). They found no performance differences for using the phone, reading or a cognitive task battery, but for mousing, sitting was best and walking was the worst and for typing sitting and standing were the best and walking the worst, with 11% fewer characters typed and 43% more errors!. So intermittent use of alternative workstations for appropriate tasks needs some planning.

As Dr Joan Vernikos, the former director of NASA's Life Sciences Division, has written "Sitting Kills, Moving Heals" (2011), we also know that exercise does not negate the hazards of extended periods of sitting (Fiorenzi, 2023). However, many offices still have just normal seating and expect workers to sit for most of the day, most don't even have sit-stand desks. In a typical day, an employee may sit to eat breakfast, sit to travel to work, sit all day at work, sit to travel home, sit to eat dinner and then sit to watch TV. Too much sitting that without variation in posture and exercising, all this sitting will be miserable and counterproductive to creativity.

1.11   WAITING TIME AT WORK

Oh dear, you are about to leave home but can't find your glasses and you need to leave now otherwise you will be too late. Aaah, where are they? Usually, you put them in a "safe" place, typically on a shelf, cupboard, by your bed or in your bathroom. That's

FIGURE 1.11

actually already too many places, of course. One consistent place is a better idea, but never mind now, where on earth are my glasses? In my inside pockets? No way. But now you urgently need them and you don't know where they are. Time passes and you get nowhere. You are starting to panic, which does not help; it only makes your searching gets more chaotic. Then a redeeming word comes from the other room: "Are you looking for your glasses"? The little voice shows some mockery. Apparently, it's entertaining to see you struggling. But you are told where you last left them and now, fortunately, your glasses have been found.

Being late and wasting time searching and waiting for things happens at work too. Usually, you have to wait before the elevator arrives, you must wait while your computer starts up. And when the computer is on, you must wait while specific software is launched. Later that day, you wait in a meeting room when other people turn up late. At lunchtime, you may wait in the queue in your restaurant. When you ask your colleagues questions, you often have to wait until you get answers from them. Before you can order something or take some action, you may have to wait for permission from your boss. On top of that, you may have a messy desk and often forget where you placed the documents that you now need. This is true for the hard-copy paper documents, and it is also true for the files on your computer, or in the company system or elsewhere on hard disks and/or USBs. In the computer, the information you need could be stored in email, chats, teams meeting notes, slide decks, documents, spreadsheets and much more. Forgetting where you stored what you need is only a problem when you need those documents or files. It is not only searching for your own files and documents but also those of others and those on the internet.

According to IBM's website (https://www.ibm.com/support/pages/cognitive-uni-versity-watson-systems-smartseller), you are not the only one searching for files. This website states:

> In 2012, McKinsey reported that "employees spend 1.8 hours every day—9.3 hours per week, on average—searching and gathering information. Put another way, businesses hire 5 employees but only 4 shows up to work; the fifth is off searching for answers, but not contributing any value". According to Interact Source, 19.8 % of business time – the equivalent of one day per working week – is wasted by employees searching for infor-mation to do their job effectively. IDC data shows that "the knowledge worker spends about 2.5 hours per day, or roughly 30% of the workday, searching for information.

A more trustful source (trustful, because it is published in the scientific literature in a way that others can repeat the same study) also shows that we lose much time search-ing for information at work. Nakash and Bouhnik (2024) distributed a questionnaire in four government offices in Israel, which was completed by 716 employees. The results show that 22% of the respondents spend about half a working day a week and 10% spend one and a half workday a week on information searches.

But people aren't all the same when it comes to information storage and retrieval. Grudin et al. (2001) found that some people are "filers" who evaluate and categorize incoming documents and file them in archives stored in cabinets that they can eas-ily access. While others are "pilers" who let papers accumulate in piles on desks or other surfaces. Pilers tend to have better spatial memories and filers have bet-ter semantic memory. They found that filers tend to accumulate more documents over time compared to pilers and store them on desktops and other surfaces. When employees had to move office, filers had an average of 20 boxes of papers before the move, and 16 after the move, while pilers had only 12 before the move and 9 after the move. They speculated that filers are eager to clean off their desks, might automati-cally even store papers that are useless, and are reluctant to throw documents away. Whereas pilers organize documents into a smaller number of often messy visual stacks and more easily toss out these stacks without sorting through them. Kwok (2015) also speculates that reducing the size of a workspace encourages a person to purge all unnecessary documents. Leanne (2024) proposes that filers may experi-ence a sense of control and order from their organized systems, but this can lead to over-accumulation, whereas pilers, despite appearing disorganized, often have a mental map of their piles that helps them retrieve documents quickly. So, systems supporting filing are useful and maybe should be used more frequently. Also, having some organization to how and where you store your stuff can help you minimize or even eliminate search time for things, and also limit your feelings of distress when you think things are "lost".

1.12 OPEN-PLAN OFFICE DISTURBANCES

Your work in the open-plan office can be disturbed terribly. One idiot can ruin the productivity of hundreds of employees in an open-plan office. You think, "that won't happen to me". Oh no? What if the windows need to be washed on the inside and

FIGURE 1.12

the window cleaner turns on his radio while working, or someone spills coffee on their brand new outfit and shouts a word that rhymes with "split", or an employee has a cardiac arrest, or someone announces very loudly that he has scored an order, or someone has a birthday and "happy birthday" is sung, or a siren draws a group of people to the window, who "ooh" and "ahh", and there is an argument about whether or not to open the windows if this can be done, and what if the program "Word" automatically adjusts the layout of your document for the umpteenth time, leading to loud primal screams from the user!!. Life can be a frustrating series of distractions! With enough employees, there is something every day, which leads to collective loss of performance. In all these cases, employees are distracted. Results from a survey of 649 workers in an open-plan office showed that disturbances and lack of privacy accounted for most of the variance in results followed by health concerns (headaches, respiratory ailments), then thermal dissatisfactions (temperature and air movement) and finally workstation features (Hedge, 1986). In a study of 779 office workers in the USA and Canada (Veitch et al., 2007), it was also found that the most annoying things in an open-plan office were dissatisfaction with privacy/acoustics, dissatisfaction with temperature/ventilation and dissatisfaction with lighting. There are many studies that indicate that an open-plan office is unsuitable for certain complex tasks. For example, Hedge (1986) and Hongisto (2005) also reported a decrease in productivity due to noise in an open-plan office, especially conversational noise because this has meaning.

These are studies where employees themselves report complaints about distractions. Whether this self-report is really reliable has been investigated by Smith-Jackson (2016). She had 17 test participants perform a web search task and a task in which a text had to be edited. She had the people perform the task quietly (without sound), with people who spoke continuously next to the person performing the task and with people who spoke occasionally. The occasional speaking did not bother them. The average performance of all participants even increased when speaking occasionally. Now it gets interesting. When you split the group of participants into people who are easily distracted and people who are not easily distracted, she saw clear differences. When navigating the web, the performance of people who were less able to focus was significantly better in the quiet condition than in both noisy environments. This trend was also visible when correcting the text, but the difference was not statistically significant. It, therefore, varies from person to person whether people can perform well in an open-plan office. For people who have difficulty concentrating, a quiet environment is more important. For the others, even hearing a few words occasionally increases performance. We should be careful generalizing this as it was a very specific study. The larger the open-plan office, the more often words are spoken, so research argues for the possibility that people can withdraw, but the open-plan office should also be limited in size to ensure that there are pauses in the conversations around employees. Medical offices in the US have to be designed so people can't accidentally overhear patient information to protect patient privacy.

But it isn't just noise and environmental control that can be a problem. Employees in a large open-plan office also breathe each other's air which can be laden with germs and pollutants, and perhaps not surprisingly, reports of "sick building syndrome" are significantly higher in open-plan offices (Burge et al., 1987; Hedge & Erickson, 1998; Hedge, 2009).

The main takeaway is that there are differences between workers in how they appreciate and perform in an open-plan office. Also, the type of task influences whether it can be done effectively in an open-plan office. If workers are handling confidential information, such as personal medical records or other confidential information, then the privacy of an enclosed office is of course required. A good option is to design an activity-based office that provides a mix of open, small group, and private offices where the nature of the work and sometimes the preference of the workers can choose the environment to do the work. In the end, an office is only an environment designed to facilitate work.

1.13   ADJUSTING THE OFFICE CHAIR

Is there is an office chair adjustment academy? There has to be; it cannot be otherwise. There are many different office chairs and the way in which you have to adjust them and the number of adjustment features can also differ for each office chair. Only very handy, intelligent, dextreous people are admitted to this academy and the students should be good at hiding the fact that they are not able to do certain tasks. This is because once you have graduated from the chair adjustment academy, you will often find that the chair you are being asked to sit on cannot be adjusted in an easy way. In that case, the graduate in question must be able to behave in such a way that it seems as if s/he has everything under control, which is, of course, not always the case. Office chairs have to meet industry standards and these say what adjustment features need to be on an office chair, especially if you want to say it is an "ergonomic chair". These standards are published by the US by organizations such as the Business + Institutional Furniture Manufacturers Association (BIFMA) and the American National Standards Institute, and internationally by the International

FIGURE 1.13

Standards Organization (ISO). Each country may have their own chair standards. But such standards typically only define dimensions and ranges of adjustment, they do not address usability concerns such as ease of use or human factors such as comfort. Office chairs also often come with an instruction manual, and typically these get torn off and disposed of by the chair installers. After all, it's only an "ergonomic" chair, so how to use it should be obvious!

When adjusting an office chair the seat height is often adjusted first. There is often a control (usually a handle or paddle) typically placed under the seat pan and on the right side of the chair. Should you pull this control up for the chair seat to go up and push it down for the chair seat to go lower. If you use it incorrectly, you may suddenly sink down, and if that happens, just when the whole department is looking at you, you will feel embarrassed. You will be sitting too low, and everyone will be looking down at you. But at a certain point, you do understand how to adjust the seat height. With pulling the lever up and simultaneously raising your buttocks up, the seat should go up and you can release the control when you are at a comfortable height. Because of the anthropometric variation in any population, the chair seat height must adjust over a range. Also, because of processes like muscle memory, your choice of what is a comfortable chair's seat height can feel different depending on whether you adjust it from low to high while sitting on it or high to low, because leg muscle activity is different in each situation. And, of course, the optimal chair seat height will also depend on the height of the worksurface you are working at. But then you also must adjust other parts as well: the seat angle, the seat pan depth of the chair, the angle of the backrest, in which ratio the backrest and chair rotate, the armrest height and depth and sometimes pivot. If you have a headrest, it is even more complicated. This headrest can also be adjusted in height and for backward position. And for every possibility of changing the chair, there is a lever, button or another way of control. And adjustments differ per chair. All these activities can benefit from training. That is why an academy is needed. After several years of study, you will have made progress and you will have learned a number of basic things reasonably well, but for the entire range of possibilities and for the entire variety of types of office chairs, a lifelong study may still not be sufficient.

Some manufacturers promote chairs with a flat seat pan, some with it contoured, others with a tilted one. Some promote additional cushioning at the side of the seat pan, some a more saddle-like form, some promote mesh seating, others different shaped cushioned seats. Some promote kneeling forwards, like the Balans chair, others promote leaning back with a full back and headrest support, like the Freedom chair. Some promote a high chair back, others a mid-height chair back, some a low chair back, and some no chair back at all! There are many different office seats available and numerous research studies on seating (e.g. Vink et al., 2007). According to Graf et al. (1995), a good chair would be a chair that supports the body in a variety of desirable positions. There is also an indication that sitting in a chair that is not static (backrest and seat pan follow the human body movements) is better for the musculoskeletal load (Robertson et al., 2022) compared with a static chair.

But with all of the chair control options and different chair designs comes considerable confusion. In the USA, one of us (the other brilliant one) investigated knowledge and use of the controls for chairs in which they sat to work. The study surveyed 1,004 randomly selected office workers from 23 different companies sitting in a total of 60 different ergonomic chairs. Apart from seat height and armrest adjustment, only a small minority of participants knew about their chair controls. Even when participants correctly identified their chair controls, less than 50% ever used the control. And surprisingly there was no effect of ergonomics training on chair comfort! However, those who had received previous general ergonomics training reported less frequent musculoskeletal discomfort than those with no training or chair-specific training. Those sitting in chairs with three or fewer controls reported significantly less frequent musculoskeletal discomfort than those in chairs with four or more controls. Overall, the results show that up to two thirds of participants knew about their controls for adjusting seat height, seat depth and armrests but most participants were unaware of controls for other functions (Hedge, 2016a).

The disadvantage of these new possibilities of supporting various postures and movements is that it becomes even more complex for the chair user. Previous studies some 15 to 25 years ago found that only a few office workers adjust their seats to the optimal position (Ong et al., 1988).

Later, in a study among office seat users, it was shown that 61% did not adjust their chairs (Vink et al., 2007). There were various arguments by the users for not adjusting the chair: e.g., it is difficult to understand the way controls work, I don't understand what is better and how to adjust my seat, usually, I don't think about the way I do sit, my health and safety adviser did put my seat in the right position, I will not change it, I can do my work, so why change the seat? The movement position (dynamic mode) of the chair is not comfortable, because there is a lack of stability.

On the other hand, Robertson et al. (2022) showed that when you train people on how to adjust the seat and explain the benefits, the effects on reducing discomfort were significant. Therefore, of course, the adjustability of chairs should be made user-friendly and consistent, but next to that training is evident. The training should be given by someone who knows the chair, the work tasks and how to prevent musculoskeletal injuries. This is all needed to prevent misery.

1.14 STOP THE MEETINGS

You are sitting for more than an hour at a table where many other employees are gathered. There you are in that room playing with your pen, looking around, not interested in the topic at all. The topic is too abstract or too detailed or you cannot follow anything of the horrible number of words that are spewed out about almost nothing. Then the chairman asks you if you do agree. You have no clue, and you play the game like the others in being vague and your answer is:

> well if I listen to the others, I hear there are several points of view regarding this point of discussion. I see advantages and disadvantages and I am not quite sure if we have enough arguments in order to make a sound decision on this point. Of course, we also have to consider the context. There are various elements at play in the context that influence the core of the matter. Most of us will of course be familiar with that. Also, because a number of things that are at play in the context now. The question is whether that will also be the case over a certain period of time. There are various forces that influence this and a good analysis of the influencing changing external factors may now be necessary. My proposal is that with a number of experts, maybe some of whom at this table may feel that they are those experts, we investigate this more deeply before

FIGURE 1.14

we can identify a number of sound directions. I deliberately speak about directions and not yet about solutions, because it seems sensible to me to indicate a direction at this stage and not go to fast into the solution mode. Of course, that also requires preliminary work. Everyone will understand that.

Only a few employees really enjoy meetings. That chatter about details that don't interest them. It's often the same people who are talking and you know in advance what some participants are going to say. It is also the question of what meetings do bring. Laker et al. (2022) report that in the USA 92% of the employees consider meetings costly and unproductive. Across the 76 companies they surveyed, employee productivity was 71% higher when meetings were reduced by 40%. It is the question of whether this is true for all companies, but we probably recognize those terrible meetings. Other empirical evidence tends to point to inefficiency as well when it comes to workplace meetings. Some studies state half of all meetings are seen as "poor" by attendees, wasting approximately $213 billion on ineffective meetings per year (Keith, 2015).

Kauffeld and Lehmann-Willenbrock (2012) in Germany showed that dysfunctional communication, such as criticizing others or complaining, showed significant negative relationships with organizational success 2.5 years after the meeting. In this study, 92 regular team meetings were studied and evaluated. Team and organizational success variables were gathered via video recordings, questionnaires and telephone interviews. The outcome of their study was also that "teams that showed more functional interaction in their meetings, in terms of problem-focused, positive, procedural, and proactive communication, were significantly more satisfied with their meetings".

We as humans decide to have meetings and we design them. There are clear indications that we have too many of those meetings and we should structure them better. The facilitator and chair of the meeting play a significant role, but the participants as well.

Apart from this, it would be great if the office design would also make these meetings much more fun. Of course, you think "but that's not because of the design, but because of the content". Yes, the content and a good facilitator play a major role in meeting quality and meeting fun (Bakker, 2016). However, meeting rooms can be designed in such a way that meetings can be a bit more fun. There are practical stories where standing meetings shorten the length of the meeting. This has even been demonstrated. Bluedorn et al. (2003) had people meet standing and sitting and it turned out that the standing meeting was 34% shorter with the same quality of results. A standing table could therefore influence the meeting. There are studies that describe the effect of lighting on meetings and there is also the idea that communication is influenced by the design. Eye contact is influenced by the design (Bohannon et al., 2013). When the PowerPoint is presented on the wall, it is difficult to gauge the reaction of other participants in the meeting, let alone respond to it. The same applies to long meeting tables, where people sit next to each other. So, in fact, the meeting environment should be chosen based on the type of meeting. In one of the other chapters, the preference for a round or oval meeting table is described, which might also have to do with the eye contact and seeing the facial expressions of others at the meeting table. This interior choice might help, but reducing the number of meetings and changing human behavior might have a larger effect on how satisfied employees will be at the meetings.

REFERENCES

Alqhtani, R. S., Ahmed, H., Alshahrani, A., Khan, A. R., & Khan, A. (2023). Effects of whole-body stretching exercise during lunch break for reducing musculoskeletal pain and physical exertion among healthcare professionals. *Medicina*, *59*(5), 910.

Baethge, A., Rigotti, T., & Roe, R. A. (2015). Just more of the same, or different? An integrative theoretical framework for the study of cumulative interruptions at work. *European Journal of Work and Organizational Psychology*, *24*, 308–323.

Bakker, I. C. (2014). *Uncovering the secrets of a productive work environment, a journey through the impact of plants and colour* (PhD dissertation), TU Delft.

Bauman, F., & Arens, E. (1996). *TaskT/ambient conditioning systems: Engineering and application guidelines. Report by University of California, Berkeley.*

Bazley, C. (2015). *Beyond comfort in built environments* (PhD dissertation), TU-Delft.

Bluedorn, A.C., Turban, D.B. & Love, M.S. (2003). The effects of stand-up and sit-down meeting formats on meeting outcomes. *Journal of Applied Psychology*, *84*, 277-285.

Bohannon, L. S., Herbert, A. M., Pelz, J. B., & Rantanen, E. M. (2013). Eye contact and video-mediated communication: A review. *Displays*, *34*(2), 177–185.

Botter, J., Burford, E. M., Commissaris, D., Könemann, R., Mastrigt, S. H. V., & Ellegast, R. P. (2013). The biomechanical and physiological effect of two dynamic workstations. In V. G. Duffy (Ed.), Digital human modeling and applications in health, safety, ergonomics, and risk management (Vol. 8026, pp. 239–249). Springer. https://doi.org/10.1007/978-3-642-39182-8_23

Bouwens, J. M., Fasulo, L., Hiemstra-van Mastrigt, S., Schultheis, U. W., Naddeo, A., & Vink, P. (2018). Effect of in-seat exercising on comfort perception of airplane passengers. *Applied Ergonomics*, *73*, 7–12.

Burge, P. S., Hedge, A., Wilson, S., Harris-Bass, J., & Robertson, A. S. (1987) Sick building syndrome: A study of 4373 office workers. *Annals of Occupational Hygiene*, *31*, 493–504.

Chung, S. (2015). *Effects of lighting on creative performance* (PhD dissertation), Cornell University. https://ecommons.cornell.edu/items/9924cae5-668b-41a1-9296-8476d7398b72?ref=gantri.ghost.io

Dazkir, S. S. (2009). Emotional effect of curvilinear vs. rectilinear forms of furniture in interior settings. Journal of Interior Design, 35(2), 55–64.

Dear, R. J., & Brager, G. S. (2002). Thermal comfort in naturally ventilated buildings: Revisions to ASHRAE Standard 55. Energy and Buildings, 34(6), 549–561.

Ellegast, R. P., Kraft, K., Groenesteijn, L., Krause, F., Berger, H., & Vink, P. (2012). Comparison of four specific dynamic office chairs with a conventional office chair: Impact upon muscle activation, physical activity and posture. *Applied Ergonomics*, *43*(2), 296–307.

Fiorenzi, R. (2023). *Sitting is the new smoking.* https://www.startstanding.org/sitting-new-smoking/

Gallup. (2015). *State of the American manager.* https://www.gallup.com/services/182138/state-american-manager.aspx

Girman Phillips, T., & Levin, H. (2009). *Critical review: How well do house plants perform as indoor air cleaners?* Proceedings Healthy Buildings, Paper 667, Syracuse University, New York.

Graf, M., Guggenbuhl, U., & Krueger, H. (1995). An assessment of seated activity and pos-
   tures at five workplaces. International Journal of Industrial Ergonomics, 15, 81–90.
Groenesteijn, L., Ellegast, R. P., Keller, K., Krause, F., Berger, H., & de Looze, M. P. (2012).
   Office task effects on comfort and body dynamics in five dynamic office chairs. Applied
   Ergonomics, 43(2), 320–328.
Grooten, W. J., Äng, B. O., Hagströmer, M., Conradsson, D., Nero, H., & Franzén, E. (2017).
   Does a dynamic chair increase office workers' movements?–Results from a combined
   laboratory and field study. Applied Ergonomics, 60, 1–11.
Grudin, J., Whittaker, S., & Hirschberg, J. (2001). The character, value, and management of
   personal paper archives. ACM Transactions on Computer-Human Interaction (TOCHI),
   8(2), 150–170. https://doi.org/10.1145/376929.376932
Hedge, A. (1982). The open-plan office: A systematic investigation of employee reactions to
   their work environment. Environment and Behavior, 14(5), 519–542.
Hedge, A. (1986). Open vs. enclosed workspaces. In J. Wineman (Ed.), Behavioral issues in
   office design (pp. 139–176). Van Nostrand Reinhold.
Hedge, A. (2009). Indoor environmental quality, health and productivity. Environmental
   Research Journal, 4(1/2). 109–117
Hedge, A. (2016a). Ergonomic workplace design for health, wellness, and productivity. CRC
   Press.
Hedge, A. (2016b). What am I sitting on? User knowledge of their chair controls. What am I
   sitting on? User knowledge of their chair controls. Proceedings of the Human Factors
   and Ergonomics Society 60th Annual Meeting, 60(1), 455–459.
Hedge, A., Burge, P. S., Wilson, A. S., & Harris-Bass, J. (1989). Work-related illness in office
   workers: A proposed model of the sick building syndrome. Environment International,
   15, 143–158.
Hedge, A., & Erickson, W. A. (1998). Indoor environment and sick building syndrome complaints in air-conditioned offices: Benchmarks for facility performance? International
   Journal of Facilities Management, 1(4), 1–8.
Hedge, A., & Lawson, B. R. (1979). Creative thinking. In W. T. Singleton (Ed.), The study of
   real skills, Vol. 2: Compliance and excellence (pp. 280–305). MTP Press.
Hedge, A., Mitchell, G. E., & McCarthy, J. (1993). Effects of a furniture-integrated breathing-
   zone filtration system on indoor air quality, sick building syndrome, productivity, and
   absenteeism. Indoor Air, 3(4), 328–336.
Hedge, A., & Ruder, M. (2003, October). Dynamic sitting – how much do we move when
   working at a computer? Proceedings of the Human Factors and Ergonomics Society
   47th Annual Meeting, Vol. 1, pp. 947–951.
Hedge, A., & Sakr, W. (2005). Workplace effects on office productivity: A macroergonomic
   framework. Cornell University, Department of Design and Environmental Analysis.
Hesselink, J. K., van Duijn, B., van Bergen, S., van Hooff, M., & Cornelissen, I. E. (2004).
   Plants enhance productivity in case of creative work. Journal of Environmental
   Psychology, 24(2), 161–170
Honan, M. (2015). Mobile work: Ergonomics in a rapidly changing work environment. Work,
   52, 289–301
Hongisto, V. (2005). A model predicting the effect of speech of varying intelligibility on work
   performance. Indoor Air, 15, 458–468.
Kauffeld, S., & Lehmann-Willenbrock, N. (2012). Meetings matter: Effects of team meetings
   on team and organizational success. Small Group Research, 43(2), 130–158.
Keith, E. (2015). (55) Million: A fresh look at the number, effectiveness, and cost of meetings
   in the US. In Lucid meetings blog. https://www.lucidmeetings.com/blog/fresh-look-at
   -meetings

Keller, A. C., Meier, L. L., Elfering, A., & Semmer, N. K. (2020). Please wait until I am done! Longitudinal effects of work interruptions on employee well-being. *Work & Stress, 34*(2), 148–167.

Kinch, L. J. (1982). The effect of furniture arrangements on the social interaction of institutionalized elderly. Environment and Behavior, 14(2), 247–259.

Kong, M., Zhanga, J.,.Danga, T. Q., Hedge, A., Teng, T., Carter, B., Chianese, C., & Khalifa, H. E. (2019) Micro-environmental control for efficient local cooling: Results from manikin and human participant tests. *Building and Environment, 160*, 106198. https://doi.org/10.1016/j.buildenv.2019.106198

Kooij-de bode, H., Blok, M., & Groenesteijn, L. (2009). Measured satisfaction in meetings at different meeting tables. In P. Vink (Ed.), Aangetoonde effecten van het kantoorinterieur (pp. 135–147). Kluwer.

Kwok, R. (2015). *To file or pile?* https://www.lastwordonnothing.com/2015/01/05/to-file-or-pile/

Laker, B., Pereira, V., Malik, A., & Soga, L. (2022). Dear Manager, you're holding too many meetings. *Harvard Business Review, 100*(5), 23.

Larsen, L., Adams, J., Deal, B., Kweon, B., & Tyler, E. (1998). Plants in the workplace: The effects of plant density on productivity, attitudes, and perceptions. Environment and Behavior, 30(3), 261–281. https://doi.org/10.1177/001391659803000301.

Lawler, E., & Hedge, A. (2007, October 1–5). *Effects of a dynamic seat pan on torso movement, back comfort, and task performance.* Proceedings of the Human Factors and Ergonomics Society 51st Annual Meeting, pp. 544–548.

Leanne. (2024). *Pilers vs. Filers - A showdown in paper organization.* https://www.containerstore.com/blog/posts/piler-vs-filer-paper-storage-blog

Lee, S. Y., & Brand, J. L. (2005). Effects of control over office workspace on perceptions of the work environment and work outcomes. *Journal of Environmental Psychology, 25*, 323–333.

Mathew, B., & Walarine, M. T. (2020). Neck pain among smartphone users: An imminent public health issue during the pandemic time. Journal of Ideas in Health, 3(special 1), 201–204.

Nakash, M., & Bouhnik, D. (2024). *How much time does the workforce spend searching for information in the "new normal"?* iConference 2024 Proceedings.

Norman, D. A. (2013). *The design of everyday things* (Revised and expanded editions ed.). The MIT Press. ISBN 978-0-262-52567-1.

O'Reilly, J. T., Hagan, P., Gots, R., & Hedge, A. (1998). *Keeping buildings healthy: How to monitor and prevent indoor environmental problems.* J. Wiley & Sons.

Ong, C. N., Koh, D., Phoon, W. O., & Low, A. (1988). Anthropometrics and display station preferences of VDU operators. Ergonomics, 31(3), 337–347.

Oppezzo, M., & Schwartz, D. L. (2014). Give your ideas some legs: The positive effect of walking on creative thinking. *Journal of Experimental Psychology: Learning, Memory, and Cognition, 40*(4), 1142–1152. https://doi.org/10.1037/a0036577.

Park, B. J., Tsunetsugu, Y., Kasetani, T., Hirano, H., Kagawa, T., Sato, M., & Miyazaki, Y. (2007). Physiological effects of shinrin-yoku (taking in the atmosphere of the forest)—using salivary cortisol and cerebral activity as indicators—. *Journal of Physiological Anthropology, 26*(2), 123–128.

Parker, K. (2023, March 30). About a third of U.S. workers who can work from home now do so all the time. *Pew Research Center.* https://www.pewresearch.org/short-reads/2023/03/30/about-a-third-of-us-workers-who-can-work-from-home-do-so-all-the-time/

Piro, S., Fiorillo, I., Anjani, S., Smulders, M., Naddeo, A., & Vink, P. (2019). Towards comfortable communication in future vehicles. *Applied Ergonomics, 78*, 210–216.

Robertson, M. M., Lee, J., Huang, Y. H., & Schleifer, L. (2022). Virtual office intervention effectiveness: A systems approach. *Work, 71*(2), 451–464.

Robertson, M. M., & Vink, P. (2012). Examining new ways of office work between the Netherlands and the USA. *Work, 41*(Suppl. 1), 5086–5090.

Smith-Jackson, T., Middlebrooks, R., Francis, J., Gray, T., Nelson, K., Steele, B., ... Watlington, C. (2016). Open plan offices as sociotechnical systems: What matters and to whom?. *Work, 54*(4), 807–823.

Sommer, R. (1969). *Personal space: The behavioral basis of design.* Prentice Hall Inc.

Statista. (2024). U.S. workers working hybrid or remote vs on-site 2019-Q2 2024. *Statista.* https://www.statista.com/statistics/1356325/hybrid-vs-remote-work-us/

Tackenberg, Y., de Wit, H., Sonneveld, M., & Vink, P. (2012). Persoonlijke inspiratie op een flexwerkplek. *Tijdschrift voor Ergonomie, 37*(3), 33–37.

Tsipursky, G. (2024, April). *The future of work will be personalized.* https://intentionalinsights.org/the-future-of-work-will-be-personalized/

US Bureau of Labor Statistics. (2023). *Telework, hiring, and vacancies news release.* https://www.bls.gov/news.release/brs1.htm

Veitch, J., Charles, K., Fraley, K., & Newsham, G. (2007). A model of satisfaction with open plan office conditions: COPE field findings. Journal of Environmental Psychology, 27(3), 177–189.

Veen S van, Hiemstra-van Mastrigt S, Kamp I, Franz M. (2012) Requirements for the back seat of a car for working while travelling. In: Vink P, ed. Advances in Social and Organizational Factors. Boca Raton: CRC Press. 731-738.

Vernikos, J. (2011). Sitting Kills, Moving Heals: How Everyday Movement Will Prevent Pain, Illness, and Early Death--and Exercise Alone Won't. Linden Publishing.

Vignais, N., Miezal, M., Bleser, G., Mura, K., Gorecky, D., & Marin, F. (2013). Innovative system for real-time ergonomic feedback in industrial manufacturing. *Applied Ergonomics, 44*(4), 566–574.

Vink, P., Porcar-Seder, R., de Pozo, Á. P., & Krause, F. (2007). Office chairs are often not adjusted by end-users. In *Proceedings of the human factors and ergonomics society annual meeting* (Vol. 51, No. 17, pp. 1015–1019). SAGE Publications.

Voss, M. W., Prakash, R. S., Erickson, K. I., et al. (2010). Plasticity of brain networks in a randomized intervention trial of exercise training in older adults. Frontiers in Aging Neuroscience. https://doi.org/10.3389/fnagi.2010.00032

Wagner, A. S., Kilincsoy, Ü., Reitmeir, M., & Vink, P. (2014). Adaptive customization–Value creation by adaptive lighting in the car interior. *Advances in Social and Organizational Factors.* AHFE Conference. pp. 40–50.

Wajcman, J., & Rose, E. (2011). Constant connectivity: Rethinking interruptions at work. *Organization Studies, 32,* 941–961.

Walsh, D. (2023). *How many Americans are really working remotely?* https://mitsloan.mit.edu/ideas-made-to-matter/how-many-americans-are-really-working-remotely

Xylakis, E., Liapis, A., & Yannakakis, G. N. (2021). *Architectural form and affect: A spatiotemporal study of arousal.* 2021 9th International Conference on Affective Computing and Intelligent Interaction (ACII). IEEE, pp. 1–8.

Yalcinkaya, G., Sengul Salik, Y., & Buker, N. (2020). The effect of calling duration on cervical joint repositioning error angle and discomfort in university students. *Work, 65*(3), 473–482.

2 Where Am I?

2.1 COMFORT IS RELATIVE

It's a nice day at the beach. However, there is always that moment that you want to go in the sea, but given that you are not on a tropical beach, you think that it may be too cold and there is a trade-off threshold for you to build the courage to take the plunge. For most of us, the sea attracts, but not the cold. A dilemma, which destroys the fun of water. Even though you have previously braved the cold, plunging into the water is still daunting. Often on the beach there is already an area with shallow water to wade in. And even this pool water is cold. So you know that the sea will be colder. But, like a "polar bear" you throw caution to the wind and run into the cold sea and move around in the water to generate body heat. On the way back out of the sea, it's interesting. Now, when you wade in the shallow water, it feels warm. That is because human sensors don't detect absolute values but rather they are good in recording differences in experiences. We don't know what the exact temperature is in the sea where we are, unless we look at a thermometer.

All our senses behave like this. Even when we decide to do something as simple as sitting in a chair. The visual appearance of the chair primes us to expect different levels of comfort – is the chair metal, wood, plastic or padded material? Does the chair look simple or complex to adjust? Is it a nice color or not? The initial position of a chair also influences the way we adjust a chair. Wang et al. (2024) showed that if we ask for the ideal seat pan angle, it is close to flat when we start with a horizontal seat pan position. When we start with a very tilted seat pan position, the preferred position is close to also being tilted. The fact that the reference (or previous value) influences sensory perception has not only been observed in chair adjustment, where proprioception plays a role, but also in temperature, noise, light, in fact, anything where other sensors are involved. The preference of the indoor temperature that people say is comfortable depends in part on the outdoor temperature (De Dear & Brager, 2002). In the northern hemisphere, office employees prefer a higher indoor temperature in the summer than in the winter. To investigate whether the influence of the previous value (the reference) also influences sitting experiences, Van Veen et al. (2016) conducted an interesting experiment. They first had people sit on a hard stool and then in a car seat, which was covered with a sheet. They then had subjects sit in a soft luxury armchair and then in the car seat covered with a sheet. Half of the 26 participants started on the stool and the other half in the armchair. Both conditions were conducted at the same time of day but on different days. The participants were told that the two car seats were subtly different. The sheet hid the difference. The story told to the test participants was that they had to get used to the temperature and environment for a few minutes and were then asked to sit down. There was only one seat (the armchair

DOI: 10.1201/9781003637035-2

FIGURE 2.1

or the stool). After the stool, the chair with the sheet was experienced as significantly softer than after the armchair, and the reverse was found for those sitting in the armchair and then the car seat. This experiment confirms that our sensors are not good at determining absolute values but can perceive differences. This also seems to work at a higher level in our brain that differences are important for the appreciation of our emotions. Frijda (1988) described the principle of "the comparative feeling" years ago: people are more positive when they are aware of a worse situation. There is also evidence that our sensors are good at perceiving fast differences in conditions and less good at slow ones, and current sensations are always influenced by the previous sensation. With this knowledge, we can therefore design better products and situations so that people experience greater comfort. For example, a lot of attention is now paid to making all phases of working in an office comfortable. The question is whether it is possible to make everything more comfortable during work and the question is whether the whole experience is then remembered as comfortable, or whether the sensory baseline also continues to shift. Is it possible to design in such a way that over time it produces moments of comfort and otherwise unnoticeable phases of less comfort. Maybe we can make cold corridors in buildings and then we will be aware of a warmer working environment even if it's just a few degrees warmer than the corridor. And for the cold sea, if we wear a wet suit initially, it will feel a little cold but quickly it will feel warmer as our body heats up the water trapped in the suit. So we all need to experience some discomfort to subsequently experience comfort.

2.2 IT SPEAKS FOR ITSELF

I work in an open-plan office writing as a journalist. For my work "silence is golden". Oh no, is that windbag of the advertising department really sitting next to me today?

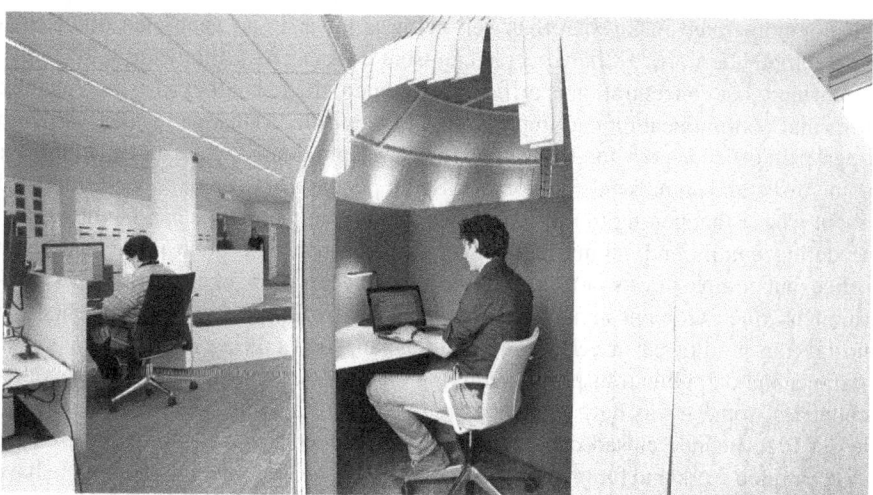

FIGURE 2.2

Why does talking while phoning have to be so enthusiastic and loud? How come there are so many people who pretend that they are funny and talk so loudly? Every phone call around me is accompanied by loud laughter. And if that was the only problem. Last time this windbag person shouted so loudly after every advertisement sold that I think blood could have dripped from my ears. His loudness was infectious. Others around me suddenly also eagerly shared their successes. At times the number of decibels became so high that it would have been more peaceful to work in a cheering football stadium.

This noise problem can occur while traveling by train, sitting an airplane or on a bus but also in any office. Of course, in the office, I like to be informed about interesting events and that works exceptionally well in our open-plan office, but at times, I also prefer some seclusion. I would rather see an overview of the number of advertisements sold at the end of the month than have to experience every success so intensely and constantly. It may sound a bit old-fashioned, but I dare to admit it on paper now. Cautiously suggesting in the cauldron of noise that I would rather quietly study a statistical analysis of sales over a week. My job is not just about following other people's sales success. I also have tasks of my own. But I am not alone.

A 1982 study of 649 employees in an open-plan office showed that 84% could easily overhear private conversations, 74% of workers reported many disturbances and distractions, 71% found it difficult to deal with confidential matters and 68% found it more difficult to concentrate. Managerial and technical workers reacted more unfavorably to office noise and the loss of acoustic privacy than clerical workers (Hedge, 1982). In a 1983 study among 100 teachers and 356 students, Becker et al. (1983) showed that people work less efficiently in an open office and that concentration is more difficult. De Croon et al. (2005) concluded, based on a literature study, that privacy and job satisfaction are low in open-plan offices. Bodin-Danielsson (2016) also demonstrated in Sweden that well-being is lower in an open-plan office than in a combination office where people can isolate themselves. Blok et al. (2012) also confirmed that concentration is difficult in an open office, but they also indicate that informal communication can improve in an open office. This means that while an open-plan office is poor for acoustic privacy and confidentiality, it can be helpful for some tasks and some kinds of workers. Some people also like to be well informed about what is happening in the company and an open-plan office can facilitate this. Updating emails and submitting invoices can probably be done in an open-plan office, but complex tasks or tasks that require concentration require a different environment. Such an environment can be created by making mutual agreements that no talking is allowed in certain areas. In Japan, loud talking is frowned upon in stations and on public transport and talking on a phone is banned on trains. In other countries, some trains have a silence area. This can also be done in offices. Even better is to include closed-off single rooms in the interior, which can be used for concentrated work and for private phone calls. The obvious solution is simply to have different workstations and to match the workstation to the demands of the task and then encourage people choose the place where their task can be carried out best. For some tasks, you need several screens and for discussions you need a room which can occupy more people. It is important for the employer and facility manager to offer these facilities. Ultimately, the office is only there to help people do their work. Make sure that the right workstations are there, and employees can perform well. And add silence areas in public transport or other public spaces.

2.3   BOREDOM DOESN'T WORK

Creativity is important in many situations. You might need to be zen to create new ideas. Complete calmness might be good. Your head gets empty, and the new ideas pop up. So, you try to come up with new ideas lying down. Just relax in a relaxed position, almost flat and some electrodes on your head capturing the brain waves and a computer will convert it into text and your boss will receive those texts. You earn your money in the easiest way. Wonderful. Doing your work lying down and the entire building's temperature adapts locally to you. The spot where you have the perfect temperature for your head, for your feet, hands and every part of the body. It is exactly the same as your desired temperature. The contact with your body is also perfect, hard on the spots where you want it firm and soft on the spots where you like that. Also, color, smell and sound is calming, the way you like it. It seems wonderful, but it's actually boring and it doesn't work. People are not suited for this. Sitskoorn (2015) describes in her book that people react less aggressively to a provocation when lying down than when sitting upright. That lying down might be good for a conversation where there is much tension between opinions of people. So, if someone wants to fight with you, just ask that person to lie down and you do it as well and the problem will be solved. So if Sitskoorn is correct, couples should never fight while lying in bed – hmmm, maybe this is a bit exaggerated. The question is also if people perform well on creativity tasks when lying down?

FIGURE 2.3

Bluedorn et al. (2003) showed that standing meetings are shorter with the same quality of outcome, probably because people are more actively involved in a meeting when standing up than when sitting down. So if you want to perform well, lying down and sitting relaxed is perhaps not completely the perfect situation.

Patterns that are the same are also not good for being creative. It might feel relaxed, having it exactly the same every day. But who still enjoys the same thing every day. If everything is within the comfort zone, then boredom may be at its peak. The boredom. That lifelessness. Light, sound, everything at one volume and at one tone. Who wants to be that super gray mouse? According to Sitskoorn (2015), it is not good for our brains to always experience the same thing. The ideal office temperature is described in many scientific studies, but the question is whether that is preferable. People who work in naturally ventilated buildings, where windows can be opened, become accustomed to thermal diversity (De Dear & Brager, 2002), and their preference for a comfortable temperature has a greater spread than that of people who always work in the same temperature. If you say it simply: people become accustomed to a greater variation in temperature and feel comfortable at more temperatures. It also does depend on age. Older people have a smaller range of a comfortable temperature (Roelofsen, 2016). Maybe this is because young people have to deal with more variation. But it is not just temperature. Being in contact with the environment outside the office is also important. Variation in view is also good. Bazley (2015) has shown that a view of the mountains in Colorado influences productivity and health. She compared employees who had this view with employees who did not. And what did she find: employees with a view performed better. They see more variation: clouds, rain, sun, dark and night, and this is more stimulating. Incidentally, the fact that you can open a window also has this effect. Even ballet performances are judged as better in a ballet theater with a view than in a room without a view (Bazley, 2015), because there is more variation (see figure). It seems nice to have so much peace and having everything the same. Especially in these hectic times, but actually we should be grateful for a bit of hecticness. It is good for performance when there are no distractions and there is a certain dullness. But this dullness should not be exaggerated. It might not help people in the long run. The brain becomes less flexible, people become more sensitive to temperature variations and inspiration is lacking, so dullness does not work. In designing offices or other spaces where people perform, it looks as if contact with the outside situation is preferable and variation in the interior might help as well. Variety is the spice of life!

2.4   PEOPLE ACCEPT MISERY

Ergonomic designs often strive to improve comfort and minimize discomfort. Yet sometimes it seems that humans look for misery. Driving in a low down sports car that is difficult to get into and out of and has a hard seat. Riding for hours on an uncomfortable bicycle saddle. Exerting ridiculous forces, lifting weights in a fitness center. Running a marathon and getting more and more aches and pains with time. People seem to like that. Somehow we are all looking for extremely uncomfortable situations. It happens. Sitting cramped in an airplane, not knowing where to put your legs eventually and struggling with your posture for a few hours. People do everything to get to their holiday destination as cheaply as possible. They sit in their car in a traffic jam for hours, they quietly stand in line at the airport like zombies, they overpack and try to take in all their belongings with them while traveling, they wait in line to hire their rental car and they buy water for expensive prices, when they have it almost for free from the tap at home or from a water fountain in the airport. And people love to tell stories about their uncomfortable experiences, about what discomforts they had to tolerate – it shows how tough they are, they find it cathartic! So, in the end for many people, comfort is not important at all. What about when at work? At the office, we might as well sit on concrete blocks with poor lighting and a lot of noise, no table, the laptop on your lap (it's called a laptop, so that's how it was originally intended, place it on top of your lap). Every ten minutes, someone comes along to give you a hard slap in the face and now and then it gets completely dark, and you get an occasional bucket of ice-cold water thrown over you. Some will like

FIGURE 2.4

it and say give me more. Air conditioning or heating is completely unnecessary, just dress appropriately and the problem is solved.

Yet, there is a difference with traveling, doing sports and temporarily sitting in a sports car. In order to perform well in races, the seat of a racing car is custom-made so that it perfectly fits the driver. Then comfort suddenly does matter, and the race doesn't last too long. The solar race car also has a seat that is made based on 3D scans of the people who have to sit in it, like it is done with the M-series of BMW car seats (Franz et al., 2011). And at the office, good performance is also required. Kosonen and Tan (2004) have shown that at a temperature of 27°C, productivity is 30% lower than at 21°C. Well, then air conditioning is, of course, worth it. Investing in air conditioning might pay off. And Hongisto (2005) saw productivity decrease by 7% when doing work which needs good concentration when there is a lot of noise. A concentration room might also pay for itself. And Derango et al. (2003) determined with productivity measurements that a very poor ergonomic workplace yields 17.5% less output than an optimal one. This research by Derango et al. (2003) is unique because productivity was actually measured. Three groups were "treated" in both the government and a company. One group received new furniture, including good seats, one group received training and new furniture, and one group was a control group and received "nothing". After a few months of work, productivity was measured. In the government, this could be measured by collected taxes over 11 months for the three groups and in the business community by the number of successfully completed telephone consultations over such a period. The telephone consultations had increased by 8.3% and the amount of collected tax by 17.7%. The training and furniture group was compared to the control group. That this training is important is also evident from research by Hedge (2016) and Robertson et al. (2009). So, sitting on concrete blocks is not the best idea. And just giving new furniture without training will also have a negligible effect. Now most workplace designs worldwide are getting better, and a 17.5% productivity increase will never be achieved simply by purchasing ergonomic furniture and that effect also will not be achieved without this training either. But it is clear that the attention to comfort and creating a good workstation can yield a lot. It is not only to satisfy workers, but it can have health benefits, prevent injuries and have some positive effects on performance even in a race car. However, comfort is not a fixed experience. So periodically experiencing discomfort and misery can help remind people of how good a comfortable place feels.

2.5 EXPERTS CAN HAVE VALUE

We all want to sit in a comfortable seat, but a chair that looks comfortable and maybe even feels comfortable at first, in the end, may not be comfortable for you. We all seek comfort. Comfort is a great subject. Theoretically, we know how to study comfort and discomfort with questionnaires and in addition, we can use measurements, like pressure distribution, motion recording, recording sound, light intensity and we use many other sensors. It is amazing what we are capable of doing research in this field. Sometimes this expert comes to you and tells you that you are sitting and moving in the wrong way. You should behave differently. The expert pretends to know

FIGURE 2.5

everything and tells you what you should be doing, and as a user you feel compelled to follow this advice. Often, you are not seen as a person by the expert, but rather as an object that should be changed and there is also the issue that every person is different – one size does not fit all! The unfortunate thing is that the expert's research knowledge may not be generalizable, because the person experiencing the situation is the only one who can say whether a product, process or system is comfortable for them, and their views can and will change over time. So a chair that initially is comfortable can feel uncomfortable after a long period of static sitting. There are always experts who think they can tell you what is good for you, but the problem is that to date no expert can predict what is comfortable over time in all situations. Of course, we might predict that sleeping on a bed of nails will be uncomfortable for most people, but depending on the spatial density and diameter of the nails we might be wrong, as many Indian fakirs might attest to. So for most products, we use it is hard to predict the comfort and satisfaction on an individual level. It partly has to do with the fact that body sizes and shapes differ (anthropometrics), but also it is affected by what the person has done beforehand (Smulders & Vink, 2021), what is their mental state, what are they used to, how long will they use the product and in what are environmental conditions they will be in. Even in a very good business class seat in an aircraft, the sensation of comfort will drop in the course of time (Smulders & Vink, 2021). There are experts who say that they know that a chair feels comfortable, but the question of whether you are sitting comfortably is something only you can answer yourself. This means that when designing a chair that has to be comfortable, it should always be tested with end users (sitters). Ideally, these are end users from the target group. A chair that is bought by a company with female employees should therefore not be tested by only a male designer sitting in it. It sounds very obvious, and you think: "is this what Vink and Hedge know so much about"? Then we have to answer you: yes, that is what Vink and Hedge know so much about. Car seats crash tests are done with male dummies and we assume that it is safe for females as well, which is debatable. In many cases, chairs are still designed without involving end users, even some business class aircraft seats lack good tests with end users. Insufficient account is taken of the task, anthropometry, emotion, behavior and culture. Yes, we come in different shapes and sizes, with different clothing, different preferences, different expectations, different durations of using the product, etc., so trying to accommodate everyone is a daunting task! It is also difficult to set up research so that data are gathered in a way that can be translated back into the design. We are capable of doing that to a certain level. For example, did you know that visual information influences comfort? The visual input gives an initial impression of comfort. Incidentally, the impression that is created on the basis of visual information can differ from the impression that is created by the feeling of sitting a while on a seat. Kuijt-Evers et al. (in Bronkhorst et al., 2001) demonstrated this difference in 2001 with 49 experienced office workers. One of the four chairs was negatively assessed on its appearance (visually). The seats were exactly the same and differed only in color. The ugly colored seat was rated less comfortable. Only after sitting for a while did the test participants find that the chair was as comfortable as the others. In an unpublished consulting study, Hedge took office chairs that were rated as old and uncomfortable and re-covered with colorful new material too thin to impact skin pressure distribution, and employees rated these new chairs as much more comfortable than their previous chairs, although they were the same chairs! Bouwens et al. (2018) tested various inflatable neck pillows in an airplane. At first glance, the pillow that was completely around the neck like a kind of collar, which is worn in case of injuries, scored the worst, but after two hours of sitting in an airplane, it turned out to be the most comfortable. The other way around is, of course, also possible because designers focus on visual appearance and often a beautiful-looking chair might feel very uncomfortable after sitting for a while. This principle can also be used to shape behaviors so if you don't want someone to sit for a while, provide them with an uncomfortable seat. For these kinds of things alone, it is very important to involve end users in various phases of the design process. Not only just watching them or doing a quick test, but systematically testing chairs with a standard protocol that allows chair comfort to be compared across different designs. When choosing a chair, it is also important that the right people (a good reflection of the population) do the test and let people work in the chair for a day and then evaluate it. So, you already know that end users have to be involved in testing designs, and doing this in the right way is something Vink and Hedge have experience in, but for your own case, only you yourself using a specific product can assess your comfort because you are your own expert.

2.6 SAVING ENERGY

In an office building, a battle often breaks out between energy consumption, comfort and productivity. Who wins? The manager who wants to save energy and is environmentally conscious should win, providing it is a cost-effective solution. In many parts of the world, large buildings are not heated in winter and workers simply are expected to wear warmer clothing, and here the scrooge, who wonders why coats have to be taken off indoors, thinks that the scarf can easily stay on, the hat on and gloves on. Hand out some second-hand wool thermal underwear as a Christmas

FIGURE 2.6

bonus and voilà: lower heating costs; fantastic! Cut the names out of your received Christmas cards and replace these by new names and they can be returned to sender immediately. Very cheap. That way, you never forget anyone; you recycle, and you are a considerate friend or employer to your staff and a responsible entrepreneur to your environment. This scrooge will do well to join the top management in saving money. People often don't remember which card they sent anyway. Oh yes, away with those energy-guzzling desktops, laptops and tablets. On smartphones, which you charge yourself by cycling at the office, you can do enough work these days. You also can get it warm while cycling, exercise generates body heat. After which you can cool down with a cup of uncooled ice tea, which also makes offering of hot drinks unnecessary. There is no need to cool the ice tea because the office you are in is already cool anyway. There is a slight disadvantage. The performance in tasks that need thinking is worse in a cold environment. In a too hot environment, the performance is worse as well. Kosonen and Tan (2004) have shown that productivity for thinking tasks is 30% lower at 27°C compared to 21°C. Hedge and Gaygen (2009) conducted a month-long field study in a US company measuring keying performance for office workers and found that when air temperature increased from 20°C to 25°C, typing errors fell 44% and typing output jumped 150% – chilly workers made more errors and cooler temperatures could increase a worker's hourly labor cost by 10%. Kim and Hong (2020) also state that a good work temperature in the office is 25°C.

So, too much air conditioning may sometimes be counterproductive. Productivity decreases at a lower temperature. The optimal temperature can differ per employee, task and outside temperature and humans differ in their clothing and preferred temperature (Roelofsen, 2016). This is also recently affirmed. Fan and Zhu (2024) showed that the effect of indoor temperature varied greatly and depended on the task type (between 17°C and 36°C). It makes sense, of course, that when you are active in the gym, another temperature is preferred than when sitting and working at your laptop. The adaptive model of thermal comfort, based on analysis of 21,000 sets of data compiled from thermal comfort field studies conducted in 160 buildings located on four continents in varied climate zones, adjusts the "comfort" temperature by season, work situation and whether a building is naturally ventilated or air-conditioned (De Dear & Brager, 1998)

Where possible a person's air temperature should actually be locally adjustable. A solution that is receiving increasing attention to optimize for a sustainable office and that can save energy is to lower the temperature in the office in the winter and install a heater at the desk or work with ventilation on the desk that blows hot and cold air or a heating element on the desktop or under the desktop. In summer, those who find it too hot can have fans and localized air-conditioning. Also, individual preferences differ. If you move from a cool climate to a warm climate, then over time, your comfort temperature will change to be warmer, and if the reverse happens it will change to be cooler. New microenvironmental solutions are being developed for offices and other communal settings (Kong et al., 2019). In addition, employees really appreciate this control over the temperature. There is sufficient literature that shows that when employees have control over the climate, the feeling of comfort is high. For optimal comfort, it is important to go a step further than a legal obligation. Although not all studies point in the same direction, there are clear indications that having control over the indoor climate has positive effects on comfort, job satisfaction and productivity (Lee & Brandt, 2005; Bordass & Leaman, 1997; ASID, 1998). Research is ongoing on the best ways to design a comfortable indoor climate (also see Chapter 1.5), but the end user having control is an important factor, which should be taken into account and has been demonstrated in studies by Roelofsen (2016) and Bazley (2015). However, after 50 years of research on design technologies to give employees personal control of work environment conditions, most companies are still reluctant to relinquish control of the work environment conditions to their employees.

2.7 REWARD

We pretend that we want to win, but there is mostly just a slight chance that you can win the competition or tournament. There is only one number one. So, we must be honest, and we will have to admit that we, in principle, love losing. In a tournament with 24 teams, only one will win. So, we love to lose. We design systems in such a way that most people lose. For example, in an international soccer tournament, 24 teams participate with say 20 people per team. In that tournament, $23 \times 20 = 460$ people will lose. And then millions of people are watching at home, and they know almost for sure that their team is going to lose as well. So, we like to lose. Why else would you sit and watch these matches? Millions of people also participate in lotteries and the chance that you win that million is almost zero. That is interesting. Football tournaments, car racing competitions, lotteries are designed in such a way that there is only one winner. That sounds nice "a winner". However, this is only for the lucky few. Most of us become losers. The systems are designed in such a way that the chance that you lose is the greatest. People apparently like that. People

FIGURE 2.7

designed these systems. Competitions for basketball, volleyball or football or any other sport are also designed in such a way that at the end of the year, there is only one winner. So, you can calmly tell everyone on the street that you are a loser. You can also do that at school, in the hospital and in the company. The new boss is telling the employees, you are all worthless. The company is doing badly. There is no chance for success. There is actually no point in what you are doing as an employee, because the chance that you will continue with the company is almost zero. Just do what you want, it doesn't matter anyway. You are all losers. So, we like misery. The negative. Or is it about that spark of hope in sports and the lottery? Psychologists have long know that intermittent reinforcement, the chance of an occasional positive experience, is a powerful way to shape behavior. So you gamble on a lottery because you hope you will win, and win big, even though the chances are very, very, very slim. In the USA, they are 1 in around 300 million, but that chance generates over $100 billion in ticket sales!

Maybe we should redesign systems to give more people more possibilities to win. Many management books are about the benefits of positive reinforcement to change behavior and to create more "winners" (for example Allen, 2006; Block, 2016; Covey, 1992). Stimulating and emphasizing the positive is much more rewarding than punishing the negative. However, in many organizations there are bad leaders (Erikson, 2021) and micromanagers, who keep track of exactly what you do, so that they seem to have control, or managers who constantly point out and emphasize the mistakes of their employees. That is frustrating and most of us have had that experience as a part of our life. Strange that it is possible that so many bad managers are still at those positions. Humans design this situation. So, we have the choice to make it better.

One of the authors (Vink) was just checking the exams of university students and his wife sat opposite placing stamps or stickers on the workpieces of primary school students who had done well. Vink took those stamps and then also gave a stamp for everything above a seven for the university students (in the Netherlands you get a grade 0–10, where 10 is excellent). It was a childish picture of a happy elephant. Funny enough, the students (20 years old or older) also went to look and said, "do you have a stamp"? "how nice is a stamp?" There was not one winner here, but about half of the group got a stamp. So, there were several winners. The university students probably liked this, because it has something to do with their primary school days. Or maybe, they just liked the picture. But the principle of rewarding good behavior is not new. Much has been written about this (for example Patel & Balady, 2020; Chen 2023). Direct positive rewards stimulate good behavior (Skinner, 1938) and help people develop a better self-image. An occasional compliment does a person good, but we should also look for how we can have more winners in systems we design. In the US, it is commonplace for all children on a team to win a medal simply for participating, and then a few get an additional medal for outstanding performance by winning games. Perhaps we should strive for more winners and thus reward more people. That is more fun, and it is be good for personal development as well.

2.8   IN LINE AT THE SUPERMARKET

At 6 o'clock in the evening, you might be standing in line at the supermarket, because apparently everyone has the idea after work that we should go to the supermarket at 6 o'clock, so we all do it. Long lines are at all the checkouts, and also at the self-scan. And we all think the same thing standing in line: "things aren't really moving fast". And then there's this calming store background music playing, which can become irritating, because you don't want calming music, you want to check out as fast as

FIGURE 2.8

you can and get home. And through the music, you can hear a few babies and children crying, because they have to come with their parent(s) to the supermarket. And when paying, the parent(s) must be partly busy with the children and partly paying attention to the cashier and partly putting the groceries on the conveyor belt and partly putting them in the shopping bag. Sharing attention and dealing with multiple tasks invariable slows a person down. You might think you can multitask, but most likely you can't and by trying to engaging in multiple tasks, your performance may suffer. And in part, that's why it takes so long at the checkout, and it is frustrating to others in line. You simply can't put the groceries on the conveyor belt, talk to the cashier, pay and put the stuff in the bag and look after your child or children. That's too many activities at the same time. Then the cashier also asks for your customer store card. Where is it again? Well, it's always in a place where you can't find it and you only get the discounts with a customer card, of course. So, you need it, and all other customers are waiting in line behind you. Couldn't you have anticipated this and had the card ready at hand? If you go to five supermarkets in the area, you will have five customer cards that you have to store and find the one you need at exactly the right moment. And you do not only have customer cards of supermarkets. Your wallet or purse is too small for this, and you need other means to put them in. It's not just the moment and searching for the card, but you also need to know the type of supermarket. Actually, you should first walk outside to know which supermarket it is, but then it's easier to ask the cashier. That's more difficult with the self-scan, of course. But then you can try all the cards, and you'll automatically see which one this supermarket takes. It sounds simple, but the people behind you look over your shoulder and see that it takes so long of course and make impatient sounds and gestures. And the customer still needs to find their credit card to pay for the groceries! By the way, in practice, this is why the self-scan checkout is appreciated by customers (Rinta-Kahila et al., 2021). Oh yes, at the regular checkout a checkout divider also has to be placed in between your and the previous customer groceries, so that the cashier knows who is the next customer. Of course, the cashier knows who is the next customer as the previous one just left but s/he has to know which groceries belong to which customers. Customers approaching the checkout lines also have to make a decisions about which line they will choose or which they are allowed to choose, Some lines are for ten items or less and others have no restrictions. But the shortest line isn't always the fastest; in part, it depends on how many items customers have and whether some items require additional work by the cashier. If you are unlucky, you may feel like you always choose the wrong line, as usually the customer before you in line forgot to weigh their vegetables and put the bar code sticker on them or they have a bar code sticker but it can't be read by the system. Then another employee has to be called and has to weigh the items and interact with the software to select the item, it's weight and the charge. Oh dear, then it takes even longer to check out. It can be so annoying because you want to checkout, get home and eat your evening meal.

Also, supermarkets have the habit of moving the products around regularly. That means you often have to search where the product is. This way shopping lasts longer. Much longer than you want. You think, just leave it in the same place, which is much more convenient. Of course, the supermarket owners want you in the store as long as possible, so that you come across as many things as possible that you want to have. You might think that product always comes in handy at some point. But after the product has been in your house for a while, it is conveniently thrown away when it is past its sell-by date. That's another thing, the past products had no sell-by date. Could that beer that is past its sell-by date really be poisonous? Did we drink in the past poisonous beer? Or is it a marketing trick to quickly buy new ones? As a customer, you don't have the time to study this for all products, so just believing the out-of-date time is easiest. At the express checkout, you might secretly count with the person standing there whether this person has more than ten groceries, because the basket seems very full. And yes, you count 13. The self-scan express checkout simply accepts that irrespective of the number of items, so why does the supermarket still have an express checkout then? This is annoying. And then, one of the biggest annoyances, the unhealthy food is always cheaper, is better packaged and is placed clearly in the field of vision, especially at the checkouts while you have time waiting in line so you may make an impulse purchase. Even if healthy products are close to the checkout, unhealthy snacks are not purchased according to Huitink et al. (2020). Fruit and vegetables on arrival gives the impression that the store is healthy, but they do not influence purchasing behavior. According to Huitink et al. (2020), the most effective way to discourage unhealthy snack purchases at supermarket checkouts is a total substitution of less healthy snacks by healthier alternatives. Checkout times can be reduced in many ways, like smart planning of the number of cashiers or for example by RFIDs on the products (Bocanegra et al. (2020). The question is, of course, whether supermarkets want this. If you wait a long time at the checkout, you might still see something you want to add, impulse buying may be more likely when you are bored (Gray Group International, 2024).

Agent based simulation software modeling can simulate a supermarkets checkout zone to compare different designs for an optimal flow. So, with newer techniques, we can optimize customer flow for cashiers at checkouts (Netlogo, 2024).

REFERENCES

Allen, T. D. (2006). Rewarding good citizens: The relationship between citizenship behaviour, gender, and organizational rewards 1. *Journal of Applied Social Psychology*, 36(1), 120–143.

ASID. (1998). *Workplace values: How employees want to work*. The American Society of Interior Designers.

Bazley, C. (2015). *Beyond comfort in built environments* (PhD thesis), TU Delft.

Becker, F. D., Gield, B., Gaylin, K., & Sayer, S. (1983). Office design in a community college effect on work and communication patterns. *Environment and Behavior*, 15(6), 699–726.

Block, P. (2016). *The empowered manager: Positive political skills at work*. John Wiley & Sons.

Blok, M. M., Groenesteijn, L., Schelvis, R., & Vink, P. (2012). New ways of working: Does flexiblity in time and location of work change work behaviour and affect business outcomes? *Work*, 41, 2605–2610.

Bluedorn, A. C., Turban, D. B., & Love, M. S. (2003). The effects of stand-up and sit-down meeting formats on meeting outcomes. *Journal of Applied Psychology*, 84, 277–285.

Bocanegra, C., Khojastepour, M. A., Arslan, M. Y., Chai, E., Rangarajan, S., & Chowdhury, K. R. (2020, September). *RFGo: A seamless self-checkout system for apparel stores using RFID*. Proceedings of the 26th Annual International Conference on Mobile Computing and Networking, pp. 1–14.

Danielsson, C. Bodin (2016). Office type's association to employees' welfare: Three studies. *Work, 54*(4), 779-790.

Bordass, W., & Leaman, A. (1997). Strategic issues in briefing, design, and operation: Future buildings and their services. Strategic considerations for designers and clients. *Building Research and Information, 25*(1), 190–195.

Bouwens, J. M., Schultheis, U. W., Hiemstra-van Mastrigt, S., & Vink, P. (2018). Expected versus experienced neck comfort. *Human Factors and Ergonomics in Manufacturing & Service Industries, 28*(1), 29–37.

Bronkhorst, R. E., Kuijt-Evers, L. F. M., Cremer, R., Rhijn, J. W. van, Krause, F., Looze, M. P. de, & Rebel, J. (2001). Emotion and comfort in cabins: Report TNO, Hoofddorp. Publ .nr. R2014871/402005.

Chen, Z. (2023). The influence of school's reward systems on students' development. *Journal of Education, Humanities and Social Sciences, 8*, 1822–1827.

Covey, S. R. (1992). *Principle centered leadership*. Simon and Schuster.

Croon, E. de, Sluiter, J., Kuijer, P., & Frings-Dresen, M. (2005). The effect of office concepts on worker health and performance: A systematic review of the literature. *Ergonomics, 48*, 119–134.

De Dear, R., & Brager, G. (1998). Developing an adaptive model of thermal comfort and preference. *ASHRAE Transactions, 104*(1), 145–167.

De Dear, R. J., & Brager, G. S. (2002). Thermal comfort in naturally ventilated buildings: Revisions to ASHRAE standard 55. *Energy and Buildings, 34*, 549–561.

DeRango, K., Amick, B. C., Robertson, M. M., Rooney, T., Moore, A., & Bazzani, L. (2003). *The productivity consequences of two ergonomic interventions*. Upjohn Institute Staff Working Paper No. WP03-95. www.upjohninst.org.

Erikson, T. (2021). *Surrounded by setbacks: Or, how to succeed when everything's gone bad*. Random House.

Fan, X., & Zhu, Y. (2024). Effects of indoor temperature on office workers' performance: An experimental study based on subjective assessments, neurobehavioral tests, and physiological measurements. *Ergonomics, 67*(4), 526–540.

Franz, M., Kamp, I., Durt, A., Kilincsoy, Ü., Bubb, H., & Vink, P. (2011). A light weight car seat shaped by human body contour. *International Journal of Human Factors Modelling and Simulation, 2*(4), 314–326.

Frijda, N. H. (1988). The laws of emotion. *American Psychologist, 43*(5), 349–358.

Gray Group Internation. (2024). *Impulse buying: Understanding and controlling spontaneous shopping*. https://www.graygroupintl.com/blog/impulse-buying

Hedge, A. (1982) The open-plan office: A systematic investigation of employee reactions to their work environment. Environment and Behavior, 14(5), 519–542. https://doi.org/10 .1177/0013916582145002

Hedge, A. (2016). *Ergonomic workplace design for health, wellness, and productivity*. CRC Press.

Hedge, A., & Gaygen, D. (2009, October 19–23). *Office environmental conditions and computer work performance*. Proceedings of the Human Factors and Ergonomics Society 53rd Annual Meeting, San Antonio, pp. 488–492.

Hongisto, V. (2005). A model predicting the effect of speech of varying intelligibility on work performance. *Indoor Air, 15*, 458–468.

Huitink, M., Poelman, M. P., Seidell, J. C., Kuijper, L. D., Hoekstsra, T., & Dijkstra, C. (2020). Can healthy checkout counters improve food purchases? Two real-life experiments in Dutch supermarkets. *International Journal of Environmental Research and Public Health, 17*(22), 8611.

Kim, H., & Hong, T. (2020). Determining the optimal set-point temperature considering both labor productivity and energy saving in an office building. *Applied Energy, 276,* 115429.

Kong, M., Zhanga, J., Danga, T. Q., Hedge, A., Teng, T., Carter, B., Chianese, C., & Khalifa, H. E. (2019). Micro-environmental control for efficient local cooling: Results from manikin and human participant tests. *Building and Environment, 160,* 106198. https://doi.org/10.1016/j.buildenv.2019.106198

Kosonen, R., & Tan, F. (2004). Assessment of productivity loss in air-conditioned buildings using PMV index. *Energy and Buildings, 36,* 987–993.

Lee, S. Y., & Brand, J. L. (2005). Effects of control over office workspace on perceptions of the work environment and work outcomes. *Journal of Environmental Psychology, 25,* 323–333.

NetLogo. (2024). *NetLogo - agent based modeling – SuperMarket.* https://github.com/terman37/NetLogo-Agent_Based_Modeling-SuperMarket

Patel, N. B., & Balady, G. J. (2010). The rewards of good behavior. *Circulation, 121*(6), 733–735.

Rinta-Kahila, T., Penttinen, E., Kumar, A., & Janakiraman, R. (2021). Customer reactions to self-checkout discontinuance. *Journal of Retailing and Consumer Services, 61,* 102498.

Robertson, M., Amick III, B. C., DeRango, K., Rooney, T., Bazzani, L., Harrist, R., & Moore, A. (2009). The effects of an office ergonomics training and chair intervention on worker knowledge, behavior and musculoskeletal risk. *Applied Ergonomics, 40*(1), 124–135.

Roelofsen, P. (2016). *Modelling relationships between a comfortable indoor environment, perception and performance change* (PhD thesis). TU Delft.

Sitskoorn, M. (2015). *IK² – De beste versie van jezelf.* Vakmedianet.

Skinner, B. F. (1938). *The behavior of organisms: An experimental analysis.* Appleton-Century.

Smulders, M., & Vink, P. (2021). Human behaviour should be recorded in (dis) comfort research. *Work, 68*(s1), S289–S294.

Veen S van, Vink P, (2016), Can Prior Experience Influence Seating Comfort Ratings? Ergonomics in Design 24: 16-20.

Wang, X., Grébonval, C., & Beillas, P. (2024). Effect of seat back angle on preferred seat pan inclination for the development of highly automated vehicles. *Ergonomics, 67*(5), 619–627.

3 Is Your Home Your Castle?

3.1 LOOK BEFORE YOU LEAP (OR SLEEP)

Look before you leap. This idiom means: analyze the situation and think about the possible bad results of your action before doing it. This is much truer than the average person thinks. Of course, you are not average, but to be sure, advertisers and salespeople want to influence behavior even when you think you don't want to be influenced. A mattress is an everyday product we use for around one third of our lives. But buying a mattress is not easy. Some people buy a mattress by laying on it in a store and then deciding. However, at home, you lie down to sleep for much longer than in the showroom and you wear different clothing. The salesman's story can influence you as well. A good salesman might know what a "good" mattress is. However, some sales stories you might hear may be complete nonsense. For example, the idea that a good mattress keeps your spine straight (like in the picture) is nonsense. There is no scientific evidence that a straight spine is better or healthier. On top of that, people often change their posture when sleeping. Skarpsno et al. (2017) show that there are on average 1.6 (SD 0.7) position shifts per hour while sleeping. De Koninck et al. (1992) even report 2.7 shifts per hour for persons between 35 and 45 years old. In all these cases, "look before you leap" is certainly true. Studying mattresses in advance is definitely recommended, and decisions can be more complex than just sleeping posture, such as how durable is the mattress, how it is constructed, are materials "natural", what does it costs, etc.

Nowadays, we talk a lot about behavioral change to help people live healthier lives. "Look before you leap" could be an important element in this. Bargh (1996) demonstrated the importance of "look before you leap" years ago. He recorded the time it takes two groups to walk to the elevator. One group had to make sentences with words about the elderly before walking to the elevator and another group had to make sentences with neutral words. What was found? The group that first thought about the elderly walked significantly slower to the elevator. Thinking about a topic influences behavior. This is also called priming. But such an effect can be dependent on the user. Doyen et al. (2012) repeated the experiment and did not find this effect. So the effect of priming may not be universal. Priming can occur partly unconsciously. Naddeo et al. (2015) had 41 people sleep on two mattresses and asked about their comfort. The mattresses were identical, but the subjects were told that mattress A was of high quality with a high price and mattress B was a cheap one with less quality. On a scale of 0–10, mattress A scored a 7.9 for comfort and mattress B a 6.4. This was a significant difference. This means that our experience of comfort is also influenced by the information we have received in advance. People's experiences can

DOI: 10.1201/9781003637035-3

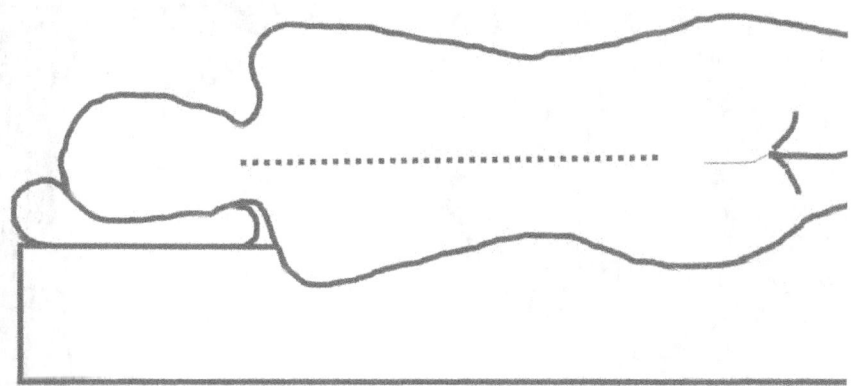

FIGURE 3.1

therefore be influenced, and advertisers know all about this! People can unknowingly prime themselves. Nobel Prize winner Kahneman (2011) states in "our fallible thinking" that there are two systems in our brains that influence our behavior. One works quite impulsively and that is why you can act quickly. It also sometimes makes incorrect decisions. For example, if an expert in a court case immediately thinks "he could be the perpetrator", the brain quietly assumes that he did it. By the way, that has nothing to do with whether or not the suspect is a bad character. Another example: after lunch, you may feel you have less energy and judges also appear to convict people more often after lunch, because making that decision may require the least effort. According to Kahneman (2011), this phenomenon occurs because the impulsive system is busy with lunch choices. The other system "thinks" more about things, and the trick is to activate that system in important situations. You may actually want to eat that high-fat, high-calorie pastry, but the other system has to activate the long-term rational thought that the size of your body has grown fast enough lately. That is the same at the office. A lunch walk would be good, but "let me just sit down for a while to rest" often wins. Also, it is good advice to do some research on a product or situation before you make a choice or take any action. For example, in choosing a good mattress, you must test it, try it for more than a few minutes dressed as you would be for bed and think about whether you will do work on it, like using a laptop on a bed, which puts you in a different posture and can increase injury risks (Bubrick & Hedge, 2014). In short, "Look before you leap" is therefore simply good advice: take the time before you act, do some research if you can, and think about what is really important for your health, your performance and your well-being, even when choosing a mattress.

3.2 REPLACING THE DUVET COVER

Just putting a clean duvet cover on should be easy. First, take the old one off, of course. But just doing that is not easy. Especially if you are doing a two-person duvet

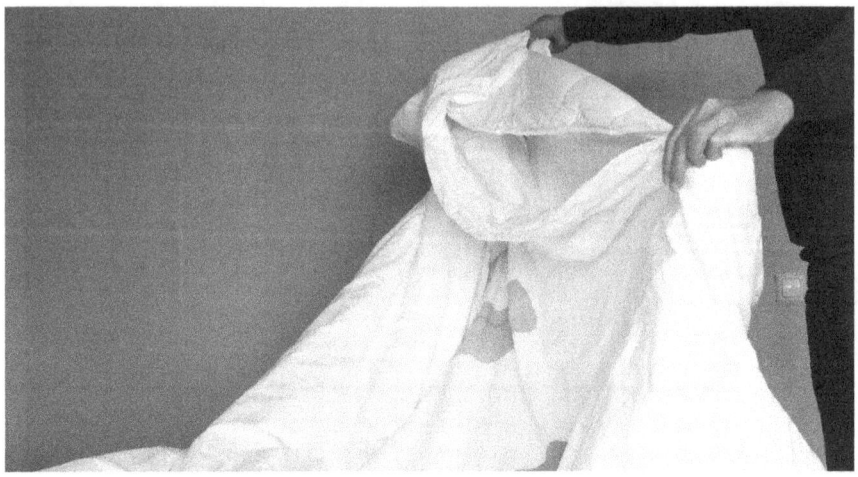

FIGURE 3.2

cover alone, it is quite a struggle to get the duvet out of its cover, but eventually, after much effort, success. However, then the bigger problem arises, putting the new cover on. Especially for anyone who is a clumsy person or has arthritic hands or some other strength and dexterity issue. Shouldn't an everyday product be made in such a way that it is easily manageable by the majority of people and not by an exclusively trained minority? For those who have blankets and/or sheets, bedmaking is easier. You simply remove the dirty sheets, then you put the clean sheets on your bed and, if you use it, the blanket over the sheets. With sheets, you only have to tuck them under the mattress, and you have a lovely clean bed in no time. But today, simply having sheets and a blanket is not sophisticated for today's lifestyle. Throwing a duvet over the bed rather than having sheets sounds simpler, but it's actually made bedmaking more complicated. It seems that duvets are deliberately designed to be less manageable for most people. But duvets have become fashionable, and as often occurs fashion and function and ease of use can be in conflict. In the master course, "how to change a duvet cover", you are taught to first turn the duvet cover inside out. Then you place your hands in the corner points. That is quite a challenge with a two-person duvet which is more than 2 m wide, since one person cannot reach that distance even with their arms wide open. You need the arm span width of a gorilla (2.6 m)! It is also physically challenging and fatiguing to stand with your hands wide apart with a duvet cover in your hands for a while. But that is beside the point. Then you have to grab the duvet by other points of the duvet. That is a challenge since you are already holding something. If you are already holding something with both hands, you cannot grab something else with both hands. So that becomes a challenge because now you have to try to grab four points with both hands. Sometimes it works by first grabbing one side with one hand. You then have a hand free to put the duvet point in the other hand. But then there are two more points: the one of the duvet cover and the one of the duvet. How do you get them on the other hand? First, look for the point in the duvet cover again and then find the duvet point by feel. Sometimes you let go with the other hand and you can start again from scratch. Putting on a duvet cover like this takes at least a few hours. Suppose you have the four points in two hands, and you have your arms wide, which is by the way one of the toughest tasks in the gym. Then you still have to shake it so that the cover falls over the duvet. And of course, it gets stuck everywhere, so that just doesn't work right away. Moreover, the length of the duvet and the cover is much too long, and it falls on the floor. And if it gets a bit dirty due to the floor, you can immediately replace the duvet cover again. But anyway, in the end, there is a cover around the duvet. And finally, in the evening you can sleep under a clean duvet cover. After a night's sleep, you might have cold shoulders, and it turns out that the top of the duvet cover is not filled and you still have to fumble to fiddle the duvet fill up. What was an easy job for one person, changing the sheets, now becomes a very challenging job for one person, or even for two people who can still struggle to get the duvet into its clean cover.

Duvets supposedly have several benefits: the duvet insert doesn't need regular washing, you can use different duvet covers to change appearance, and they supposedly trap body heat but wick moisture away which may improve sleep quality. However, current duvet and cover designs highlight a real design error. Apart from being fashionably "Scandinavian", the reason for making this new design remains unclear as the sheet and the blanket work well, and in a warm summer, you can only remove the blanket and that is not possible with a duvet. In the Northern USA, where winters can be harsh, a comforter paired with blankets is still the most popular bedding option. To date, there has been only limited research on sleepwear and bedding materials and different designs, but what has been done has found that for young adults linen bedsheets can improve sleep quality when it is warm and goose down-filled duvets increase slow-wave sleep when it is cool conditions (Li et al., 2024).

However, research on how different bedding options affect ease of use and bed-making is lacking. Hotels sometimes don't use duvets because of the housekeeping time required to remove and replace them each day. The design of duvets highlights the importance of user research and design for ease. The importance of generating user insights during the early stages of product design has been stated before (e.g. Rohrbeck & Gemünden, 2011; Kristensson & Magnusson, 2010;). A test with a duvet cover change would have shown some issues. On the other hand, there is much more misery on everyday products use than the duvet cover and it is probably hard to prevent all cases in which the misery occurs.

3.3 OPENING FOOD PACKAGING

The management and designers of the packaging industry need to be tackled vigorously. Opening packaging often is a misery. It can be so difficult and can take so long that before you can even get the pre-packed product open; by the time you can open the packaging of your slices of cheese, they are already going moldy! Product adverts sometimes say, "in handy take-away packaging". By the way, is regular packaging that difficult to take along. No one informed us of this, but maybe that's beside the point. The question then arises: wouldn't all other packaging be portable? Is all the

FIGURE 3.3   Opening food packaging

other stuff we buy not transportable? Have we missed something? If you buy a container of orange juice and drink some, then you can still take the package with you. But what about food packaging of say cheese or sausage. The packaging often says, "open here" at one corner. That corner looks suspiciously like the other corners. With such a corner, the trick is already to separate the two layers of plastic. Sharp nails are a must! But even then you need excellent vision or you might need to use a microscope to see where the separation is between the layers. Suppose you have discovered the separation between the layers, then the trick is to get a layer between your thumb and index finger. With a slightly thick finger or a damp or greasy finger, it won't work anyway. I also don't have a knife or scissors at hand. Should I eat it with the plastic and all? How do you get to the slice of cheese? In any case, I have no grip. The friction coefficient between my finger and the plastic is too different from what should happen functionally. Who comes up with these corners on the cheese packaging? They are small, smooth and are too rigid, which means that opening them requires a separate investigative study. Is there a master's degree in "opening packaging"? In any case, the designer seems to have little knowledge of user research. By the way, this is not something specific to a country. This problem occurs in many places in the world. There is also already scientific research that has studied this phenomenon.

Ideally, packaging has to satisfy four sets of user requirements:

1. Safety – packaging has to ensure product integrity, maintain its freshness and eliminate any contamination, and much current packing does this. However, packaging should also be designed to include consideration of user safety concerns. Some packaging has to be designed to be difficult to open, for example, to prevent a child from opening a pack of pharmaceutical pills that could cause illness for them.
2. ReUse – unless it is a single-use product, packaging should be easily resealable, like what happens with say a screw top jar.

3. User diversity – packaging should be easy to open by a wide range of users, young and old, one and two-handed, those with limited dexterity e.g. arthritic fingers, those with visual difficulties, those wearing say gloves.
4. Conspicuity – it should be easy for the user to know where the product opening is for the packaging in a range of conditions e.g., different lighting, different temperatures, dry and wet conditions, etc.

Winder et al. (2002) report that 67,000 people in the UK visit hospital casualty departments every year due to accidents involving food or drink packaging. They performed a study among 200 people and asked them what packaging causes problems. Fifty-five percent were not strong enough to open glass bottle jars and 21% were not strong enough to open flexible packaging. The main reasons for problems in opening packages are the large forces required, tear tabs that are too small and the poor visibility of the opening mechanisms (Dittrich & Spanner-Ulmer, 2010; Marks et al., 2012). Often people think that they are not smart or strong enough to open the packaging, but it is important to mention that it is not the person using it who is doing it wrong, it is the packaging industry, that needs to study how people can open their products more easily.

3.4   REMOTE CONTROL

What a great idea. You don't have to stand up from your seat anymore to control things. You can put the television, music, refrigerator or turn any electronic device on and off, and adapt it to your wishes by pushing a few buttons. The remote control seems an ideal product. Even with your smart phone, which can act as a remote control, you can control everything. So, lets put the television on with the remote control. If you have an old remote control, it could be that the signs on the buttons have disappeared from wear. That is really annoying. The previous user might have

FIGURE 3.4

remembered the icons on the buttons, but a new one will not know this. For private televisions, this is not an issue, but in the hotel, at work or in teaching and changing to another room, this is serious. Trying to set guidelines to unify this is probably impossible. We are not even able to make the power socket the same worldwide. For a new remote control, the on and off button is easy to find. The icon is a circle with a horizontal line crossing the top of the circle. At least that is what I had in mind. So you can push the button and…oh, nothing happens. So you push it again and again nothing happens. Maybe the battery in the remote control is dead. Let's change the battery. Oh no indication of how you open the remote, should it be sliding, pressing, pulling or…? Just try different scenarios and perhaps after losing a nail, sweaty armpits or irritation you throw it on the ground – that opens the battery area! So, fortunately, you had the correct size batteries in house and you changed the batteries and tried it again. Pushing the on/off button did not work again. You try pushing it again and walking around then suddenly the television comes on. So, the button does work, but it takes a while to do this or maybe there is a range or direction issue. This is annoying. This should be designed completely differently.

There are some basic principles in designing products for users. One of them is for the designer to use International Standards for the icons on hardware and software products (ISO 7000, 11581, 13251. Another is to provide immediate feedback (Norman, 2006). It is important for a user to get immediate feedback otherwise the user gets lost. In many systems, it takes a while before the device or software works. Feedback is essential, otherwise the user does not know what is happening. There are two feedback forms: activational feedback and behavioral feedback. The activational feedback is important. The moment you press a button you see that there is activation of some function or when pressing the button there could, for instance, be a light blinking on the television or your computer screen with a spinning circle showing that the operation is running. You then know something is going to happen. The behavioral feedback means that the television is turned on or that the software is operational or the affirmation that the email has been sent.

The misery exists because designers, computer programmers or engineers sometimes forget that there is a user working with their system. This user needs to be supported in working with systems. The user needs information and feedback when the user tries to activate something. This begins at the start. Often, there is the time needed to get systems working, in these cases activational feedback is really advised. Another recommendation is to design remote controls with icons that will not be eroded and disappear over time.

3.5   THE KNIFE WITH JAM

Usually, the handle of a knife is heavier than the blade. This means that when you transport the plate with the knife on it, there is a good chance that the knife will fall off the plate. This usually happens at an inconvenient moment. For example, when walking from the table to the kitchen or vice versa. Gravity does its work and the knife falls to the ground. After a meal, the knife is, of course, often dirty. There can be jam on it or other sticky food residue. Fatty butter, olive oil or salad dressings are

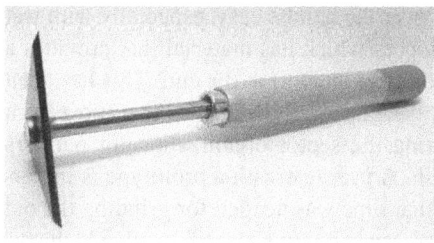 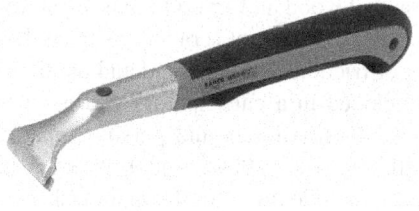

FIGURE 3.5

also other wonderful substances to fall on your beautiful carpet. On the way to the kitchen, the knife falls, and the peanut butter finds its way between the hairs of the carpet. If you have a wooden floor or tiles, the damage is manageable, but it usually happens when you have a carpet or a piece of carpet on the floor. Even when you have a wooden floor or tiles, the knife waits before falling until it is directly above your carpet. It is also possible that a whole sandwich or other food residue is dragged along in this action. Anything that sticks to the knife and anything accompanying the knife, like a sandwich, will end up on the floor. Although the odds of a sandwich falling with the bread surface down versus the butter/jammed surface should be 50/50, they aren't! Usually, the sandwich falls on the ground with the side where the jam or peanut butter is located hitting the carpet. The substance that you do not want on your carpet falls, of course, on the carpet. The bread is then no longer suitable for consumption. Well, it could be less inconvenient if the falling sandwich was still fit for consumption but it seems that this typically isn't the case. Falling sandwiches inevitably obey Murphy's law! (anything that can go wrong will go wrong).

The cause of this problem lies in the balance of the handle of the tool – the knife. When the handle is heavier than the blade and the knife lies on the edge of the plate, the heaviest part goes down first. When clearing a table after a meal usually, the knife is pushed forward a little so that the handle lies on the edge of the plate. It is not wise to place the transition from handle to blade on the plate edge as its falling down is guaranteed, and a falling knife blade could cause an injury. Placing the knife correctly on a plate also is no guarantee for success, as the knife can slide during transport. That is certainly possible when someone is walking with the plate in their hand with the knife on it as this person moves. There is an edge and the guard between the blade and the handle, but it is positioned in such a way that it cannot slide inwards to the center of the plate, but it can slide outwards from the plate. Once it is outside the plate edge, sliding back to the plate center is no longer an option and the heaviest knife part is outside the plate, and this is the preparation phase for the knife falling on the ground. It is clear that a "tool" must be designed in such a way that it fulfills the task well, but the secondary tasks of cleaning up must also be able to be done pleasantly.

Other tools have design issues as well. For example, a paint scraper had a triangular shape and a tapered handle. With a thick outline at the front and a thin one at the back. So, you have little grip to make the pulling movement. The handle was also made of wood and smooth, making sliding over the handle easy, especially with wet hands. A new paint scraper has been developed, which has material that provides a lot of friction between hand and handle and offers support at the end. This has been developed in a participatory process with designers, engineers, human movement specialists, painters and persons representing the sector organization for painters (Eikhout et al., 2004). And also tested again. A user test with a prototype is important. The side task (not the main task) was that time was needed for grinding the old scraper. Therefore, the new paint scraper was made out of sintered metal, which was hard enough that grinding was not needed anymore. Ergonomists, like Hedge (1998), have conducted extensive research on hand tool design and reported on the design principles that allow any hand-operated product to be used in a safe, healthy posture to sustain comfortable and productive work. This shows that hand tools can be improved and maybe the misery with the knife can also be reduced by designing it in a participatory way with input from experts and end users and testing the prototype.

3.6   REPLACING THE DOOR HANDLE

Maybe your door handles do not look good anymore. Maybe they are damaged, or they don't shine anymore, and they look no longer nice and modern. So, you just have to buy new doorknobs. Have a good look in the DIY store and yes, you've found some nice ones. You check with the shop assistant whether there are different sizes. But the assistant reassures us that there is only one common size for door handles, so you can safely take your choice home. Then at home, the work starts. First, you remove the old doorknob. That takes some effort, but it comes off. Oh, but what do you see now? Your door has been painted. Where the cover plate of the old doorknob used to be, the paint of the door is damaged. That is the case with all the doors. And

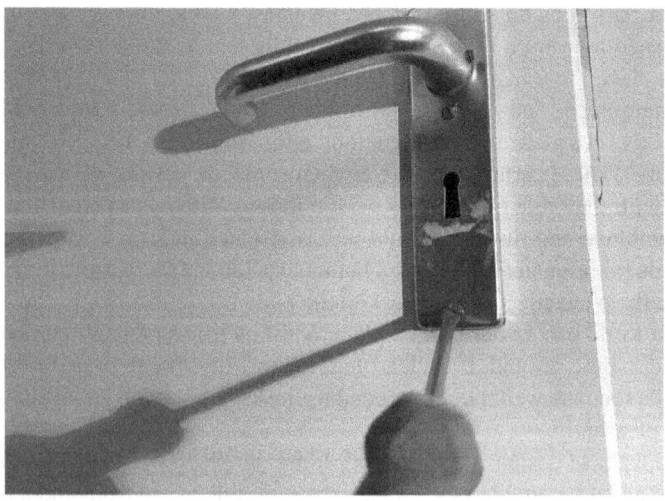

FIGURE 3.6

also the new cover plate seems a bit smaller. When you hold it against the door, yes it is smaller. If you put that on, then you will see those ugly damaged paint spots on the door. So back to the DIY store to check if there are doorknobs with a larger cover plate. Fortunately, there are. And the store manager lets you exchange the doorknobs and cover plates. So, like a happy child, whistling while going home and happy go to the doors in the house, and feeling proud that you produced this solution yourself, you think you are brilliant. It is so good to have the feeling that you are really good at doing construction jobs in the house. What a wonderful feeling to be able to solve this so well by yourself. So, you will not have to plaster, fill and paint, just replace each doorknob and its cover. Back home, you take the doorknob out of the packaging, which is a kind of hassle. But with a sharp knife and a lot of fiddling you finally succeed. However, the paper with the description and instructions was on the packaging and is torn apart now, so I have to put it together like a puzzle to be able to read it. Thankfully this is a really good cover plate for the doorknob because it is big enough. Then you discover that under the outer cover plate, there is a smaller plate with screws to attach the doorknob. The screws must be placed in the existing holes, but those holes in the door do not correspond with the holes in the purchased doorknob cover plate. So what now. The store assistant told me there is only one size of doorknob, which is true, but it turns out there are different placements for attaching screws. What a mess, the rotating part of the doorknob does fit in. Fortunately, the size of the square rod that has to be slid into the door is the right size. But the position of the screws is completely different and you have to fix that somehow. New holes have to be drilled but they will be very close to the existing ones, so that is actually not possible because then it might become a big hole. What now? Fortunately, you know someone who is a carpenter. You can just ask him. You have read that carpenter's spatial skills are very well developed, especially in experienced carpenters (Cuendet et al., 2014). So you feel sure that your carpenter friend will offer you a simple solution. But instead, his answer is:

> yes friend, your house is 25 years old and in those 25 years the sizes of doorknobs changed and while modern doorknobs are all the same size the attachment mechanisms for modern doorknobs is different. Didn't you think of this before your bought new doorknobs?

Well, no, you had not thought of that, and it turns out that when you asked the neighbors if they have experienced the same problems, they have. The designers of new doorknobs apparently have not taken into account the existing or previous sizes and versions of doorknobs. Perhaps they think: "Let us design a new doorknob from scratch. That is much more fun, then we can design much more freely". But if these designers in question had done a simple test on an old door or asked users to replace the doorknob of a previous house, they might have been able to make some adjustments to the design to make it fit. But the new doorknob does fit a new door, so maybe the designers also hoped you would replace the door as well as the doorknob – built-in obsolescence is better economics for the construction business! For new construction work, this new doorknob and cover plate are not a problem, but misery arises when replacing the doorknobs of existing old-style doors.

3.7   THE REAR LIGHT OF THE BICYCLE

You just want to turn on the rear light of your bike for the first time. It should be simple but that isn't that easy. First, you have to find the on-off button and that is a challenge because you are in the dark because you don't turn on your rear light in daylight. Finding the on-off button has to be done by touch, because it's dark. But the button is not on the top, so maybe it is on the side. No, not either. Maybe it is underneath. But you don't feel anything there either. Is it on your handlebars? No, not on the handlebar. So you feel again on the bottom of the lamp, a place that you cannot easily see, and there is a ribbed structure in the shape of a circle and when you press it, something happens. The red light starts to flicker. But you don't want that. You want constant light. What use is a light that keeps flickering, going on and off. Why would someone come up with an idea like that? Is it a pointless function? Research suggests a benefit for flickering light. Although reaction times to a flickering light are slower than those to a continuous light (Prabu Kumar et al., 2020), Gil et al. (2024) have shown that for continuous viewing a flickering light tends to be more conspicuous and attention-grabbing compared to continuous light, making it useful in situations where quick detection is crucial. So maybe a flickering light on a bicycle makes it more conspicuous at night and improves safety. But Gil's research was done on rats, so you aren't sure it works well for humans and so you look to find a continuous light.

When you press the button again, the light flashes even faster. Third time lucky? After pressing the button again the rear light switches on as a continuous light. That is what you want. How fantastic, after searching and fumbling with the button you have what you want. However, the designer of this rear light apparently never thought

FIGURE 3.7

about ease of use when the lamp is actually attached to the bicycle. Every time a user wants to turn on the continuous rear light, the user has to reach underneath the lamp and press the button three times, and maybe each time the user has to do this s/he mumbles "why does it have to be so inconvenient, so awkward". The on/off button should be visible and at a logical position (Norman, 2006). It would also be nice if the function is selectable from the handlebars, but it isn't.

As in other chapters dealing with software development, this lamp design makes it unclear what is happening in the system (Kortum, 2022). The designer may not have thought about the main task the user wants to be able to do, which is to turn on the rear light while on their bicycle. But the designer has given the user three options with no guidance on which is most appropriate for what situation, and they have not made it easy to activate these options while sitting on the bicycle.

3.8 THE RAIN SHOWER

When the shower control is placed directly under the shower head, turning on the shower tap to prevent cold water from spraying all over one's body should be done with an outstretched arm. And yes, that is one of the authors, naked in the picture, now don't get too excited! But this shower has a rain shower head and our smelly author wants a rain shower. Feeling the water falling on all places on his body. Think about it, the water is not spraying out from one smaller area, a shower head, but falling like raindrops from a large shower head that has a huge area, and the water flows over his body, which gives him a wonderful feeling and the water is felt at different places on his body. He wants that rain shower!

So, like one of the authors, you decided to have a new rain shower installed in your bathroom. After installation, the technician tested it and everything worked,

FIGURE 3.8

the water went hot and cold, so it was fine. Thanks to the technician, your enjoyment can begin. However, the first time is quite a shock. You stood underneath the rain shower, turned on the hot tap and a torrent of ice-cold water raindrops fell over you. What a hassle. And with your shower plumbing, you have to stand underneath the rain shower to turn on the tap for the water. So, now you have to experience these annoying cold water droplets every time. You think, why isn't the tap positioned so that you can reach it without standing underneath the shower head? Everyone can understand that there is a distance between the place where the water is made warm and the rain shower head in the system. Even hot water in the supply pipe will cool down when the shower is turned off, so, there will always be cold water first. You now know it now as well. However, neither the technician who installed the shower nor the person who sold the rain shower to you in the store told you about this and they could have made a plan to design it in the right way so this didn't happen. Now you have to let a cold puddle of water spray over you every time you want to take a shower. So you have now learned to reach out to the shower control with one arm, like a crane. It is a bit annoying and strange that you have to do that. You get a cold arm, but that is acceptable. Not ideal of course, but it works.

Reachability of the shower control is very important. In designing any system of controls and displays, the reach envelope of the person should be considered to optimize the design (Yang & Abdel-Malek, 2008). For instance, in designing the flight deck of an aircraft, it is important to ensure that the pilot can reach all essential controls and buttons and see all important screens from their seated position. A literature overview of available studies in this field has been made to calculate the optimal position of the controls for pilots (Joslin, 2022). These should be reachable, of course. This takes into account the variations in the human body sizes. For instance, the eye height while seated varies between persons, the torso dimensions vary, and the arm length varies as well. This arm length should also be measured in the position the hand can control the buttons. So, it is not just measuring the arm length in a standard position. The flight deck should be designed in such a way that all controls and buttons are accessible by pilots with varied body dimensions. This is critical to ensure that all tasks can be effectively accomplished in a timely manner without excessive workload (Joslin, 2022). Of course, if the reachability is not good in the airplane the consequences are much more serious than when you can't reach the shower faucets, but the principle is the same that the design should be made in such a way that humans can reach the controls or buttons. There are many ways to do this. It can be done by just testing a prototype with different-sized humans. But nowadays many digital human models are available. In those models, simulated humans can be placed in the to be designed situation and the designer or engineer can see whether those humans can reach the controls. By ensuring reachability, much misery can be prevented.

REFERENCES

Bargh, J. A., Chen, M., & Burrows, L. (1996). Automaticity of social behavior: Direct effects of trait construct and stereotype-activation on action. Journal of Personality and Social Psychology, 71(2), 230–244.

Bubric, K., Hedge, A. (2014, October 28–31). Postural risks associated with laptop use on a bed. Proceedings of the Human Factors and Ergonomics Society 58th Annual Meeting, pp. 586–590.

Cuendet, S., Dehler-Zufferey, J., Arn, C., Bumbacher, E., & Dillenbourg, P. (2014). A study of carpenter apprentices' spatial skills. *Empirical Research in Vocational Education and Training*, 6, 1–16.

De Koninck, J., Lorrain, D., & Gagnon, P. (1992). Sleep positions and position shifts in five age groups: An ontogenetic picture. *Sleep*, 15(2), 143–149.

Dittrich, F., & Spanner-Ulmer, B. (2010). Easy opening. *DLG-lebensmittel*, 5, 32–36.

Doyen, S., Klein, O., Pichon, C. L., & Cleeremans, A. (2012). Behavioral priming: It's all in the mind, but whose mind?. *PLoS One*, 7(1), e29081.

Eikhout, S. M., Bronkhorst, R. E., Vink, P., & van der Grinten, M. P. (2004). Toward a comfortable paint scraper. In *Comfort and design* (pp. 199–210). CRC Press

Gil, R., Valente, M., & Shemesh, N. (2024). Rat superior colliculus encodes the transition between static and dynamic vision modes. Nature Communications, 15, 849. https://doi.org/10.1038/s41467-024-44934-8

Hedge, A. (1998). Design of hand-operated devices. In N. Stanton (Ed.), Human factors in consumer products (pp. 203-222). Taylor & Francis.

ISO 7000. (2019). *Graphical symbols for use on equipment – registered symbols*. International Standards Organization

ISO/IEC. (2010). ISO/IEC 11581-10: Information technology — User interface icons. ISO/IEC.

ISO/IEC 13251. (2019). *Information technology — Collection of graphical symbols for office equipment*. ISO/IEC.

Joslin, R. (2022). Anthropometry considerations in the design and evaluation of flight deck displays and controls: Literature review. International Journal of Aviation, Aeronautics, and Aerospace, 9(3), 1.

Kahneman, D. (2011). *Thinking, fast and slow*. Farrar, Straus and Giroux.

Kortum, P. (2022). Where's my jetpack? waiting for the revolution in statistical analysis software interfaces, but going in the wrong direction. *Interactions*, 29(5), 68–71.

Kristensson, P., & Magnusson, P. R. (2010). Tuning users' innovativeness during ideation. Creativity and Innovation Management, 19(2), 147–159.

Li, X., Halaki, M., & Chow, C. M. (2024). How do sleepwear and bedding fibre types affect sleep quality: A systematic review. Journal of Sleep Research, 33(4), e14217. doi:10.1111/jsr.14217. Online ahead of print.

Marks, M., Muoth, C., Goldhahn, J., Liebmann, A., Schreib, I., Schindele, S. F., ... Vlieland, T. P. V. (2012). Packaging—a problem for patients with hand disorders? A cross-sectional study on the forces applied to packaging tear tabs. *Journal of Hand Therapy*, 25(4), 387–396.

Naddeo, A., Cappetti, N., Califano, R., & Vallone, M. (2015). The role of expectation in comfort perception: The mattresses' evaluation experience. *Procedia Manufacturing*, 3, 4784–4791.

Norman, D. (2006). *The design of everyday things: Revised and expanded edition*. Basic Books.

Prabu Kumar, A., Omprakash, A., Kuppusamy, M., et al. (2020). How does cognitive function measured by the reaction time and critical flicker fusion frequency correlate with the academic performance of students? BMC Medical Education, 20, 507. https://doi.org/10.1186/s12909-020-02416-7

Rohrbeck, R., & Gemünden, H. G. (2011). Corporate foresight: Its three roles in enhancing the innovation capacity of a firm. Technological Forecasting and Social Change, 78(2), 231–243.

Skarpsno, E. S., Mork, P. J., Nilsen, T. I. L., & Holtermann, A. (2017). Sleep positions and nocturnal body movements based on free-living accelerometer recordings: Association with demographics, lifestyle, and insomnia symptoms. *Nature and Science of Sleep, 9,* 267–275.

Winder, B., Ridgway, K., Nelson, A., & Baldwin, J. (2002). Food and drink packaging: Who is complaining and who should be complaining. *Applied Ergonomics, 33*(5), 433–438.

Yang, J., & Abdel-Malek, K. (2008). Human reach envelope and zone differentiation for ergonomic design. *Human Factors and Ergonomics in Manufacturing, 19*(1), 15–34.

4 Do You Really Love Your Computer?

4.1 LOGIN

You just want to order a book. This looks easy when you visit an attractive page at a website. The book is in stock. So, you immediately put the book in your electronic basket, and you are going to buy it. But wait, if you want to buy it, you have to use an account. Do you have an account? No, you don't think so. Then you'll create one. It should be easy. You just type in your email at the place where the website asks you to. But when you do this it tells you that this email is already in use. You don't recall creating an account but maybe it just slipped your mind, and apparently, you already have an account. But oh dear, you don't remember the password for that account? You have a multitude of passwords because that's what is advised: "don't use the same password for different sites". However, your memory is limited, and putting them in your computer or saving them in a browser doesn't seem very safe either. You have to log in for everything first. For your bank, for your travel insurance, planning a car service at the garage, for all the shops where you order something, for a ticket at the cinema. Of course, you run out of passwords, and remembering them all is completely impossible. And easy passwords often are not allowed. You thought about using the password "thisIdidforget", but it has to contain other characters and numbers and letters. This is specially invented so that it is almost impossible to remember. That is exactly what you don't want. You do want to remember it. That is what you want. When ordering, you first look at your mobile phone or the computer screen for a while and think. What was the password again? There is a strange thing. The word "password" is actually completely wrong, it is actually a "holdback word". A word that does not allow you to go further. The best thing is to just immediately press "forgot password" and then receive an email with instructions. But sometimes such emails are redirected into "spam" or even deleted as not being recognized, so sometimes you get nothing and then you are hopelessly stuck. But thankfully it worked this time! A new password and you have an account. So, now you have access to the shopping basket, and you can order and pay. You want to pay with your credit card. But to do this, you have to enter your credit card information and enter a secure code for the credit card. This code is another thing to remember. What was the code again? You think you have saved it somewhere. Yes, you found it after a long search. But by now, you have timed out and you have been thrown out of the website. So you have to search for the book all over again. Log in again, and do the procedure with "forgot password" again. So, you enter "forgot password" again and go through the whole process. Then you come to the secure code and now you type it again. Yes, just type it in and I'll be done, you think. But not so fast. Oh no, the

DOI: 10.1201/9781003637035-4

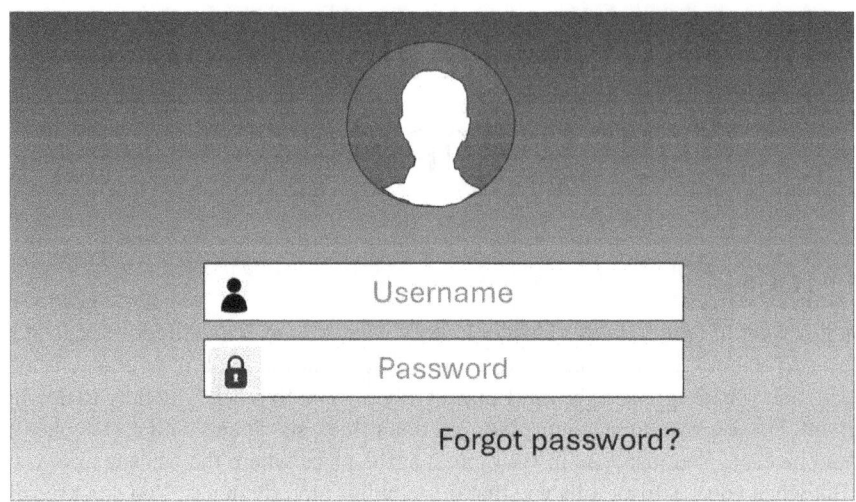

FIGURE 4.1

book is suddenly no longer in stock. The website tells me that it can be ordered if you just click "yes", which you do only to see that the delivery time is 4–6 weeks! You would have preferred to buy the book now at another store, but that is no longer possible because you have already paid at this site and you can't cancel the order. What a hassle!

Much of this kind of fiddling is due to making the system safe from hackers who can data like steal credit card details, but safety is not the only element. There is also the user. Apparently, for the designers of this website, there was a lack of user research. Did anyone simply test the system? Oh, and any testing should not be done by website builders or by people who are invited to test in a computer community as they are trained computer users and not average users. There is also a fundamental flaw in this system, where we assume that people can remember passwords that consist of special characters, upper and lower case letters and numbers. Additionally, they should be at least eight characters long. That is one more than seven, while research has shown this can pose a problem for many users. What a silly mistake. Eight is not really desirable.

Already in 1956, Miller (1956) stated that there is an upper limit of seven plus or minus two elements on our capacity to process and store information on simultaneously interacting elements. Later Saaty and Ozdemir (2003) affirmed this maximum of seven. So, the choice of eight is really silly. However, Miller also showed that "chunking" can improve information storage and recall. So in countries like the USA, phone numbers can comprise 11 or more digits are used, but "chunked"; so there is a country code "X" that can be up to seven digits, a "state" code that is three digits and "area" code that is three digits, and then the phone number that is four digits. The same chunking logic can apply to passwords but apparently, website designers are unaware of this ergonomic research!

4.2   SOFTWARE JUST DOES WHAT IT WANTS

What is happening now? You had formatted your text but now it changes automatically to something you don't want. Aligning in the left margin has been automatically changed. And it is not clear why your software did the change and now the document alignment is incorrect. If you delete your latest typed characters (undo your typing), then at a certain moment, the formatting goes back to the situation you wanted. But if you add text and tabs or paste by Ctrl V, then once again, the whole layout is disrupted, including a change in the font type. Even "paste special" does not help.

Now suppose your car behaved in the same. You are driving on the road and suddenly the car starts braking and drives itself into the verge. A terrible bump, of course. If you are lucky, it is only a bump. Your car can also tilt over and then you suddenly find yourself sitting in the car in a different way. The sky is to your left and the street is to your right. You are hanging to the side in a ridiculous position and people are looking at you. It would feel very unpleasant. Something really needs to be done to restore the situation to normal, but what. That solution isn't obvious and it cannot be done in a simple way. The car needs to be towed away and the damage repaired. As we move to more automated vehicles, such incidents of the car behaving

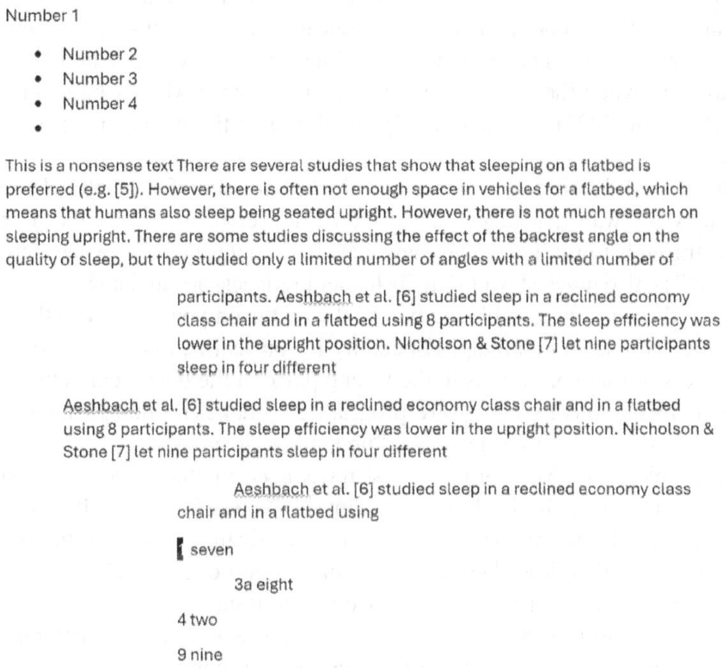

Number 1

- Number 2
- Number 3
- Number 4
-

This is a nonsense text There are several studies that show that sleeping on a flatbed is preferred (e.g. [5]). However, there is often not enough space in vehicles for a flatbed, which means that humans also sleep being seated upright. However, there is not much research on sleeping upright. There are some studies discussing the effect of the backrest angle on the quality of sleep, but they studied only a limited number of angles with a limited number of

participants. Aeshbach et al. [6] studied sleep in a reclined economy class chair and in a flatbed using 8 participants. The sleep efficiency was lower in the upright position. Nicholson & Stone [7] let nine participants sleep in four different

Aeshbach et al. [6] studied sleep in a reclined economy class chair and in a flatbed using 8 participants. The sleep efficiency was lower in the upright position. Nicholson & Stone [7] let nine participants sleep in four different

Aeshbach et al. [6] studied sleep in a reclined economy class chair and in a flatbed using

seven

3a eight

4 two

9 nine

FIGURE 4.2

autonomously' might occur more frequently. Such incidents are reported on the Tesla Motors Club website, for example, one driver said:

> On one occasion in one of our Model 3s I've had something like this happen, but at very low speed it was crawling traffic. I was probably doing around 5 mph and without any warning the car just slammed on the brakes so violently I thought I'd hit something in the road I hadn't seen. I've experienced nothing like it before or since. I've experienced AEB a few times at higher speeds (>20mph, 30kph) but this felt more aggressive and violent. Can't explain it and I just put it down to a one-off random event which I hope doesn't happen again.

There are even reports of automated cars spontaneously locking the occupants in or locking them out. The world of reliable automated vehicles isn't quite here yet.

With software glitches, the result is both physical and mental misery. And all too often when software gets upgraded, there is a new layout that requires relearning which is also very uncomfortable.

This autonomous frustration can all be prevented when the system is designed in such a way that something only changes when the user explicitly gives indicates that they want it to change. At least a button, a key or an icon should be touched by the user to initiate the change. Perhaps the intention of the design is not bad, and the designer may have assumed that the user will be happy with this automatic change, but user research should always be conducted to confirm that the change will be appreciated by the majority of the users. But today, most software is complex and has too many features and possibilities, which makes it complicated for the user and confusing when the software automatically changes what is happening in the system (Kortum, 2022). Kortum (2022) saw the same thing happening in statistical software.

With the proliferation of features and the use of icons for graphical user interfaces, many software packages have become slower to use when the user has to navigate through too many menus and have too many menu choices. Often, this "bloatware" is also much slower than its leaner predecessors, and now when you type something, there is a noticeable lag before the text is on your screen or the function is executed. This is so annoying! Especially if you realize that you have typed the wrong letters or that your text is in the wrong place in the document! The growth in features and functions in a software program can make life more complex for the user and actually slow rather than speed up their performance.

For example, when MSWord was first introduced in 1983 it could do basic text editing, formatting, spell checking and printing. Now with 2020 MS Word 365, it has become an AI-powered writing assistant with advanced collaboration features.

As Kortum (2022) describes, a lot still has to be done to make software more usable. Taking account of the user perspective in designing software is, of course, not easy as "the user" is an abstraction, and studying on what all different people want will result in a too large a list of features, as is the complexity of many software programs today (Sunstein, 2020). Maybe we have a plethora of features to satisfy all users, but ideally, we should try to find out what is the main purpose of the software for specific kinds of people and applications and ensure that we can design core capabilities that can run on a simple PC desktop, laptop or tablet, and if people want more functionality, then give them the option to activate additional features and capabilities. And above all, what is very important is "don't let the system do things by itself, that cannot be influenced by the user".

4.3   THE AI CALL CENTER

You are visiting Europe and you need help so you call the helpline number you've been given. You say "Hello, is there anyone I can talk to?" You hear a reply, Just type one for Russian, two for Romanian, three for Greek.... And at number nine, there is English. The system can see your phone number. So, it should know that you aren't local and it should be able to change the order of the language options more conveniently and put English first, but it doesn't. Never mind, you click on nine. Then you hear the next menu. If it is an emergency press one, if you are a doctor press two and if you are a registered patient press three. Here we go again. You aren't a registered patient, but you need some medical help so, Okay, you press three. Then you hear the next menu: are you a patient of this location, press one, are you a patient of another location, press two. Oh, you don't really know what locations are being referred to, so you just press one. Then you are asked to enter your date of birth. You enter 04231957 (month, day, year – the US system). But then the system tells you that this is an invalid date of birth. Now you are sure that you were born on this date so you enter it again. But the date of birth still is not valid, the system has another way of wanting to have the date of birth but it doesn't tell you what this is. So now you are getting irritated but you need some help so you try again. But what now, you were really born on that date. Oh, wait maybe you should type 57 instead of 1957. Oh no,

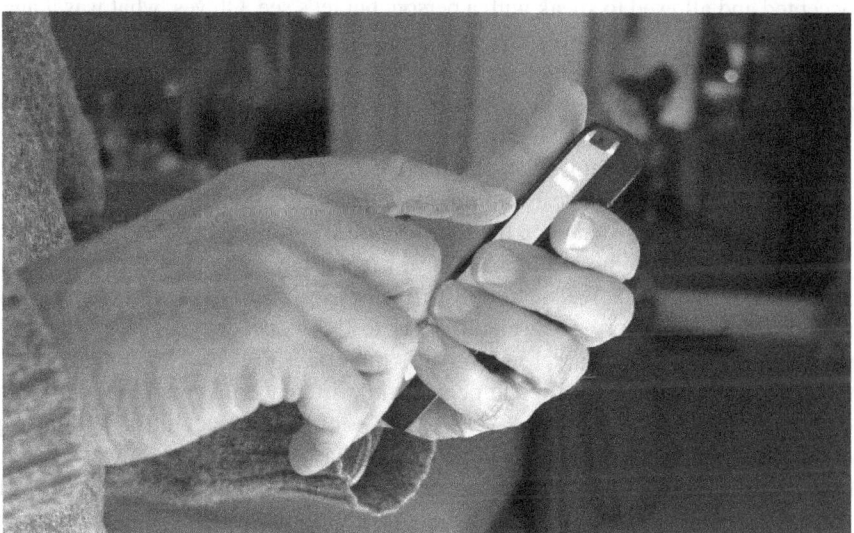

FIGURE 4.3

again this is an invalid date of birth. Instead of telling you how to do it, you simply get the answer that it is invalid. That does not help much. Oh, wait sometimes you have to type the day first and then the month, like in the English system. And yes, that works. Then you are asked to "enter your name". But wait a minute, you only have numbers on the keyboard of your mobile phone. How do you do that again? My goodness, what a hassle. Oh, wait maybe you can use voice input, maybe you can say it out loud and the system records it. So you say your name. And yes something happens. You hear "Can you say your name again and clearly". You answer saying "yes I can". Then you get "this is an invalid name". I think this is a valid answer to the question of whether you can say your name. The system should have said "say your name again, the system does not understand this", but entering into a discussion with such a system seems hopeless. You are not going to try either so you just say your name, in this case you think you should say your surname first, so you say "Vink". Then you hear "Is the name 'Pink' correct"? You say "no", but now another way of inputting your name is needed. The text on the menu appears again. Type one if the name is correct, type two if you want to try again. Okay, you type two and say your name. Then you hear: you can say your name after the beep. Then you hear "Is the name "Vink" correct"? You say "yes", but now you are asked for another input choice. You type one and then you hear "there are four people waiting before you". What is this again? A menu like this should be there so that you are helped quickly. The system doesn't tell you how long the waiting time will be and it doesn't keep you updated on how many people are waiting ahead of you. So you are patient. But after ten minutes, you decide to hang up and try again in an hour. And then you have to go through the entire menu again. Nine for the language, three that you are a patient, two that you belong to this location, at least you think so. This does not feel good as well, because you feel insecure in your answers and you just hope that you will be accepted and allowed to speak with a person, but let's see. Oh yes, what was it again with the date of birth, did you have to do the year first? What a hassle. Then you are told there is only one person before you in the system. Finally, you get through and get someone on the phone. When you complain to them about this system, you are told that there are more complaints, but that the person on the phone cannot do anything about it, it is company policy. By the way, you think this is not company policy alone. It is also those who designed this system.

There is much knowledge on the user-friendliness of these systems (Liang et al., 2024) and the mental models humans have when using a phone (e.g. Hyyppä et al., 2000; Lee et al., 2020; Weichbroth, 2020). However, theory is not enough and the application of the theory confirmed by user testing is needed as every situation is different, with different target groups. It is important to do a user study of a new technology design and let different people with different backgrounds and different cases check whether the system works properly for them. A cognitive walkthrough could also be helpful to redesign in an earlier phase by using several cases, which could reduce the burden to many of end-user studies (Farzandipour et al., 2021). Additionally, consistency of interaction is also important. A mix between input with numbers and input by voice is complex. Ideally, one type of input should used to simplify user interactions and improve user satisfaction.

4.4   THE AUTOMATIC UPDATE

You are exactly on time for the online meeting. Your electronic calendar reminds you of this. Just click on the link and it will work. At least that is what you had in mind. But no, your system thinks differently. It thinks that this is the ideal moment for an update. It seems to always do an update when you really need the system. That is what the system does and that is what the user does not want. It is fascinating what the designers of these systems have in their minds. They probably think, never update when it is best for the user, but always do that when they really need to have the system work. So, now you are too late in the meeting as the update takes time. Updating always seems to happen when it is inconvenient for the user. The designers of these update systems really like to punish users. This not only happens with online meeting systems but also with other software systems. Sometimes updates will install automatically, and often at inconvenient times. This is really terrible especially when you are in the middle of writing a document or doing data analysis in Excel or whatever. Sometimes you get the message in the middle of the work you want to finish "the update will be installed automatically now or in an hour". It now seems as if the end user has been considered but is it really true? First of all, the end users are already being disturbed in their work. This message can be planned to appear at a much better time and that can be done nowadays using AI, but probably AI is not used for this case, where it would be very handy. In addition, the choice is too limited, now or an hour is something you must think about. Stop the updates when I work, you think. Sometimes it says that you can choose a convenient time for your device to restart and complete the update, but you are already disturbed by the message. Also annoying is when you want to go home or quickly go to another meeting and you turn off your laptop. Then you get to see that the laptop should not be

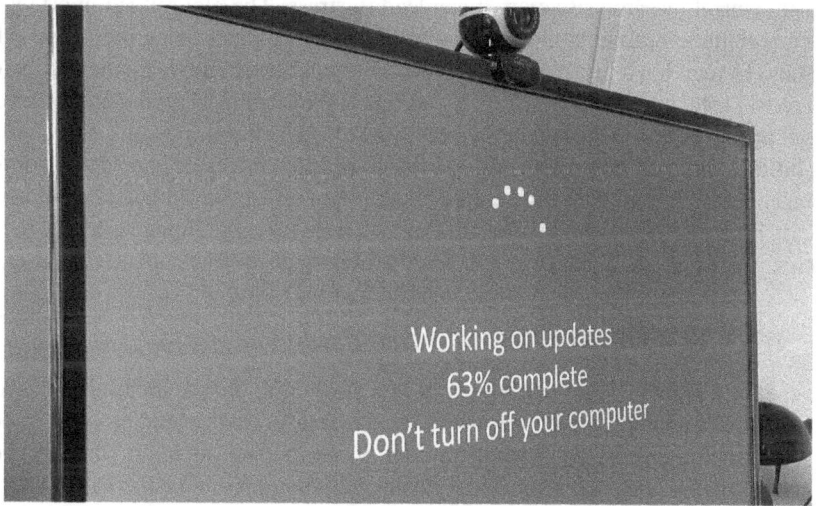

FIGURE 4.4

shut down because there are updates. Then you must leave, but you can't, you must let the laptop upload the updates first. One of the authors (Hedge) was in the back of a taxi in a country with very slow internet access shortly before a major presentation when his laptop decided to update the operating system! How stressful!! And it isn't just computers that are affected, mobile devices see more frequent updates than computers, as 38% of apps are updated at least once per week, and these data are from 2023 (ITonDemand, 2013)! It really is a disaster to have to deal with this kind of hassle all the time. I assume that designers have the best intentions and there are entire conferences about human-computer interaction, but it doesn't seem to help. It remains somewhat makeshift. One author (Vink) once thought he was being smart by using his computer completely offline. Then no virus can get in and it is safe, but then the stupid software says again that you need an update before you can continue. It is built into the operating system that you need updates, while if you work offline there should be no need for an update.

The solution is quite simple, of course. Limit the number of updates (sometimes you get an update for my phone and then there is only a button changed, which is annoying as well as you are used to the old situation), differentiate between critical and cosmetic updates, and if the update is really needed allow the user to select a good moment in time for them. All update alerts should say how important they are, how long the update will take and give the user an option to easily reschedule when the update can happen.

The problem is described in the scientific literature as well. Mark et al. (2008) describe that it takes on average more than 23 minutes to fully recover from a distraction and Yin et al. (2018) describe the technology overload by mobile applications, information, communication, and interruption. They also state that these problems are widespread in the digital workplace making work more complex while not contributing to the quality or quantity of work. Karasek (2004) found that if humans have no control over the situation, it can lead to stress. Therefore, it is important to reduce automatic updates and find good moments for updates using the latest technologies. To mention a few possibilities: we can record heart rate from the face color and arousal level by skin conduction. By adding text (or number) typing analysis, you can get an indication of how intensive the work is. Of course, there are many other possibilities and with AI it should be possible to see how intensive people work or for instance, check the calendar for a coffee break. Then select the right update moment. At least it can be done much better than it is arranged now. Humans design these systems. So, we are in control of reducing the misery of millions of software users.

4.5   TOO MUCH DATA

It is so easy nowadays to take pictures. When your child does something nice, you can take a photo using your phone, which is almost always with you, or when something else nice happens, you can just picture it. At the supermarket, you can photograph a wine bottle then send it to your friend who is a wine expert and ask "is this the good wine"? On your phone, tablet and/or computer you will probably have photos from work, from presentations, from the sunsets, from movies, from a live

FIGURE 4.5

performance by a band of your friends where they are on holiday or when something nice happens with their children. Grandparents can easily receive movies of their grandchildren and on holiday you take the same photo eight times because you want to be sure that there is one good one. Later, you think, you will sort out which one is the nicest and delete the others, but you seldom take the time to do this. Electronic memory is cheap and also your photos can easily be automatically sent to the cloud. So, there is no restriction on storage capacity. You can do this endlessly and take as many photos and videos as many as you want. You are afraid you might miss a picture of something important. So, you grab your mobile phone and you can easily take pictures and videos. In 2020, Google Photos alone had an estimated 4 trillion pictures, so these days there must be a quadrillion photos in the cloud, and it keeps growing. But for you and your photos, there comes a time when you should sort them all out. You don't want to erase everything because then your best photos will be lost. But with the huge number of images you have, selectively sorting them could take you a long time and a lot of effort and so, being a lazy human, it is questionable whether you will ever do this. Maybe there is a way to sort and filter them? With every photo you take or receive, you now start to get nervous and think, what am I going to do with that later? When will you ever have time to isolate yourself and focus on sorting, coding, filtering, clustering and deleting your photos? You may bite the bullet and start to do this, but by probably halfway through doing this you come to the conclusion that you need to label and organize images differently and now you have to start it over again. Oh no, is that your future? Will you have to neglect your children, family and friends to have time for this? If you hide away to do this work uninterrupted then, by the time you are finished, no one will know you and, having lost track of so much time, you will feel alienated from the world. Of course, the same thing happens with hard copy documents and data that you keep and file. Keep everything. You never know when you might need it. One author (Vink) once got rid of a database and what happened, just a month later, of course, someone asked for it. The other author (Hedge) replaced his work computer, going from a Mac to a PC, and the Mac had survey data on it from a lighting research study. At the time Mac files and PC software files weren't interchangeable. Well, two years later he received a request for the survey data, but by then it was long gone. Maybe this can happen again after sorting out the pictures or documents. Maybe we can keep and store electronic information endlessly, although with time software and file formats change so maybe these will become electronic ancient languages to be decrypted by digital archaeologists!

So typically, we will make hard copy printouts of our work and for data on scientific studies, and archive this hard copy so that it is traceable for ten years. Such archives can be really valuable. Research data from the famous Hawthorne lighting research studies was supposedly destroyed at the end of the research. However, one author (Hedge) found the researcher's personal copy of the data from some of the studies in an archive at Cornell University and he and his graduate students were able to statistically analyze the results to show that the original conclusions of "no effect" were incorrect (Izawa et al., 2011).

Electronic storage is a bad idea in terms of sustainability, because these data centers keep growing and they require a lot of energy, cooling water and space. In 2022, data centers worldwide consumed as much as 340 terawatt-hours of electricity which is around 1.3% of global electricity usage, and this is growing annually (Data centers & networks – IEA). Data centers now use so much energy that if you make them energy-efficient you can reduce the world's energy consumption by at least 1%. "Green modular data centers can reduce this carbon by close to 1% of total energy use worldwide" (Yang et al., 2023). But there is also an opportunity to reduce the demand for data centers. Reducing demand by designing a selective system that does not automatically save everything in the cloud, giving you alerts and warnings if you use too much data space. The latter is difficult because then you are asking companies to earn less as they earn money when you store your data in their systems. But users can also be more aware and store less. That is not only useful for energy, but finding things later may also be easier. These systems, the choices and making documents, data and pictures are all designed by humans. Of course, it is nice to have memories and data stored. It is important to archive historically significant documents. But for our personal use, we could be more selective, and we could have systems helping us to automatically sift through personal photos and files and reduce digital storage space. But we humans are lazy. We create this misery ourselves.

REFERENCES

Data Centres & Networks. (n.d.). International Energy Agency (IEA). Retrieved March 29, 2025, from https://www.iea.org/energy-system/buildings/data-centres-and-data-transmission-networks

Farzandipour, M., Nabovati, E., Tadayon, H., & Jabali, M. S. (2021). Usability evaluation of a nursing information system by applying cognitive walkthrough method. *International Journal of Medical Informatics*, *152*, 104459.

Hyyppä, K., Tamminen, S., Hautala, I., & Repokari, L. (2000). *The effect of mental model guiding user's action in mobile phone answering situations*. Electronic Proceedings from The 1st Nordic Conference on Computer Human Interaction.

ITonDemand. (2013). *The hidden value of software updates*. https://itondemand.com/2023/10/13/the-hidden-value-of-software-updates/

Izawa, M. R., French, M. D., & Hedge, A. (2011). Shining new light on the hawthorne illumination experiments. *Human Factors*, *53*(5), 528–547.

Karasek, R. A. (2004). An analysis of 19 international case studies of stress prevention through work reorganization using the demand/control model. *Bulletin of Science, Technology & Society*, *24*(5), 446–456.

Kortum, P. (2022). Where's my jetpack? waiting for the revolution in statistical analysis software interfaces but going in the wrong direction. *Interactions*, *29*(5), 68–71.

Lee, C., Kim, S., Han, D., Yang, H., Park, Y. W., Kwon, B. C., & Ko, S. (2020, April). *GUIComp: A GUI design assistant with real-time, multi-faceted feedback*. Proceedings of the 2020 CHI Conference on Human Factors in Computing Systems, pp. 1–13.

Liang, J. T., Yang, C., & Myers, B. A. (2024). *A large-scale survey on the usability of ai programming assistants: Successes and challenges*. Proceedings of the 46th IEEE/ACM International Conference on Software Engineering, pp. 1–13.

Mark, G., Gudith, D., & Klocke, U. (2008, April). *The cost of interrupted work: More speed and stress*. Proceedings of the SIGCHI conference on Human Factors in Computing Systems, pp. 107–110.

Miller, G. A. (1956). The magical number seven, plus or minus two: Some limits on our capacity for processing information. Psychological Review, 63(2), 81–97.

Saaty, T. L., & Ozdemir, M. S. (2003). Why the magic number seven plus or minus two. *Mathematical and Computer Modelling*, *38*(3–4), 233–244.

Sunstein, C. R. (2020). *Too much information: Understanding what you don't want to know*. MIT Press.

Weichbroth, P. (2020). Usability of mobile applications: A systematic literature study. *IEEE Access*, *8*, 55563–55577.

Yang, C. W., Galkin, N., & Vyatkin, V. (2023, October). *Towards interoperability of edge datacentre in the energy community with IEC 61850 modelling.* IECON 2023-49th Annual Conference of the IEEE Industrial Electronics Society. IEEE, pp. 1–6.

Yin, P., Ou, C. X. J., Davison, R. M., & Wu, J. (2018). Coping with mobile technology overload in the workplace. *Internet Research, 28*(5), 1189–1212.

5 Ouch Why Does It Hurt?

5.1 SIT AND STAND WITH THE RIGHT CONTROLS

"My field of attention increased. I can concentrate better. When I was blocked, I could easily walk away from my workstation, while I continued to fumble around while sitting". That is a nice story about standing work by Jamis (http://www.37signals.com). So, we should all do our work standing. No need to buy seats anymore. How wonderful would that be. No need to give instructions on how a seat should be adjusted. This is amazing, lower costs for the company and at home we don't need a couch anymore, we get a lot of space. Only a table is needed. All the chairs and seats can be thrown away. And for airplanes, on long-haul routes we can all stand. The literature also shows the positive effects of working standing. Sengupta and Das (2000) have shown that the reach ranges are greater when standing compared to sitting and more force can be produced when standing (Yates and Karwowski, 1992). However, standing all day is not good either. A review of 17 studies (McCulloch, 2002) shows that when standing for most of the day, there is a greater chance of blood vessel problems in the legs and problems in the lower back and feet. Alternating between standing and sitting does have beneficial effects. Aaras et al. (2001) showed that during computer work the m. trapezius (shoulder/neck muscle) has lower muscle tension during the alternation and according to Vellinga (2001), well-being increased in 84% of office workers when standing and sitting were alternated. Other studies have also shown the positive effects of the sit-stand table (e.g. Hedge & Ray, 2004). So, we should not stand all day. Apart from health issues, it is questionable whether it is comfortable to work all day standing without moving around and whether performance is improved. Sitting for long periods also is not good either. And so now we can work at a sit-stand table, which can sometimes be set to the standing position and sometimes to the sitting position. But when purchasing a sit-stand table, make sure that it adjusts high enough for comfortable standing and low enough for comfortable sitting, and that it is easy and quick to operate.

Having easily adjustable and obvious furniture controls are important. In a US survey of 1,004 office workers, 96.2% reported that their chair had seat pan height control, and 87.5% said their chair had armrest height controls, but for other control the level of awareness of the control was only between 25%–40% (Hedge, 2016).

Groenesteijn (2015) showed in her dissertation how important the right buttons are in furniture. She compared chairs with buttons that were easy to operate with somewhat more complicated systems. The adjustment time was shorter when the controls were better indicated and more intuitively located. It was also experienced as more pleasant. This research hypothesized that it is more important for flexible workers to have easily adjustable chairs than for permanent employees, because they

DOI: 10.1201/9781003637035-5

FIGURE 5.1

have to adjust them more often. Because another person has sat on the chair before (Groenesteijn, 2015). The opposite turned out to be true! The flexible workers had more experience in adjustment because they were used to using different chairs and could easily find their way around. Two types of employees, namely employees who use flexible workplaces and employees with a fixed workplace, were observed and asked about their adjustment behavior with two types of chairs. This research took place in their own work environment and with their own activities. Employees with flexible workplaces adjusted the chairs more often, also for short-term use, and were faster with a number of settings compared to users of a fixed workplace.

So, it is important to pay attention to ease of operation when purchasing a sit-stand table. One with a manual crank, which you have to turn for a while before it reaches the desired height (or depth), probably is not going to work. It takes too long and that means that people will not use those positions. If you have to crank part of your working time, that is a waste of time, and it is questionable whether productivity is promoted. Even an electrically adjustable table eventually may be set to one position, especially if it takes a long time to adjust. Too bad about the designer's effort, but it has effectively become a fixed-height standing desk. And then there are also sit-stand tables, which are just not high enough for standing work, especially for a tall person. Then the taller people stand with vulture necks to do their work. These kinds of mistakes ensure that even some people on this planet think that sit-stand tables do not work. These same people will probably also walk around a department with sit-stand tables and say that hardly anyone uses the standing posi-tion. But Vellinga (2001) showed that even if the table is used standing only 8% of the time, it still has a significant beneficial effect. In short, that sit-stand table should be easy to adjust and should adjust over the appropriate height range so that it can promote variation in posture and give space to personal preference. Paying attention to the correct operation and height is critical. But in addition to sitting and standing, periodic moving and stretching are also beneficial, and an optimal regimen seems to be 20 minutes sitting, 8 minutes standing and 2 minutes moving to be repeated throughout the day (Kar & Hedge, 2020, 2021). With mobile technology supporting us in managing our activities, there is a greater opportunity for making tomorrow's work more postural mobile.

5.2   NECK PAIN CAUSED BY NOISE

Screaming customers at an airline counter. That happens. We can understand it when people start screaming when the flight they should be on suddenly is canceled, but it is annoying, the screaming, is it necessary? Not only might the counter employee have to wipe saliva from their forehead, but the others behind and around the counter are also distracted from their work and cannot concentrate anymore. If the hearing of the counter employee adapts to the shrieking noise level then in theory the counter employee should not be able to clearly hear and understand the next customers. The loudest human scream ever recorded was 129 dB (anything over 90dB can damage hearing). So if the screamer was emitting enough decibels the counter employee's hearing could be damaged. Noise also causes problems in offices, which reduces productivity. Hedge (1982) found that 79% of workers found it difficult to concen-trate because of conversational noise in open-plan offices. Students working in a noisy simulated office environment (high noise: 51 LAeq and low noise: 39 LAeq)

FIGURE 5.2

remembered fewer words, rated themselves as more tired, and were less motivated (Jahncke et al., 2011). Evans et al. (2007) found that office noise in an open office produces more adrenaline in the blood compared to a quiet office, people are less inclined to solve problems, and the workplace is less well-adjusted, which results in poorer postures. But it is not just the loudness of the noise but also its meaning, conversational noise is more distracting than unintelligible noise. And it's not just here on earth, but astronauts riding the space shuttle and orbiting are exposed to mid-deck noise of 63dBA, and they report that it causes fatigue, difficulty sleeping, poor communication within the shuttle, temporary threshold shifts in hearing, and interference with relaxation (Segal & Hedge, 1999).

Attention to the acoustic environment is important to performance and health because exposure to noise can cause neck complaints and even give you swollen feet. This sounds completely illogical, of course. You would rather expect some damage to the ear or effects of stress, but neck complaints due to noise sound completely absurd. Yet Mellert et al. (2008) found this when people who worked in noisy aircraft were compared with people who worked in quieter aircraft. The people were not aware of the difference in noise, but the amount of neck complaints was significantly higher in the noisy aircraft. In the noisy aircraft, the employees also noticed that their feet were more swollen. Foot swelling was also the case in the other aircraft with less noise, but it was not noticed. These are strange phenomena and they appear to be the same for men and women. We only know a bit of what sound does to us. Perhaps the airline counter employee immediately looked at their swollen feet after being exposed to shouting, but we don't know that. The variety of effects of noise exposure also makes research difficult in this area, sometimes employees cannot properly assess whether they are bothered by the noise or by something else and additional measurements are needed. But if noise increases stress, and stress can increase muscle tension, then neck complaints can arise from noise at work. So there are good reasons to pay attention to acoustics in the design and during the use of an office. We should also educate people that shouting doesn't improve information processing or reduce the misery caused by noise.

REFERENCES

Aaras, A., Horgen, G., Bjorset, H., Ro, O. & Walsoe H. (2001). Musculoskeletal, Visual and Psychosocial Stress in VDU Operators before and after Multidisciplinary Ergonomic Interventions. A 6-year Prospective Study –Part II. *Applied Ergonomics, 32*, 557-559.

Evans, G. W., & Wener, R. E. (2007). Crowding and personal space invasion on the train: Please don't make me sit in the middle. Journal of Environmental Psychology, 27(1), 90–94.

Groenesteijn, L. (2015). Seat design in the context of knowledge work (PhD thesis). TU-Delft, Delft University of Technology.

Hedge, A. (1982). The open-plan office: A systematic investigation of employee reactions to their work environment. Environment and Behavior, 14(5), 519–542.

Hedge, A. (2016, September 19–23). *What am I sitting on? User knowledge of their chair controls.* Proceedings of the Human Factors and Ergonomics Society 60th Annual Meeting, Vol. 60, pp. 455–459.

Hedge, A., & Ray, E. J. (2004). *Effects of an electronic height-adjustable worksurface on self-assessed musculoskeletal discomfort and productivity among computer workers.* Proceedings of the Human Factors and Ergonomics Society 48th Annual Meeting. HFES, pp. 1091–1095.

Jahncke, H., Hygge, S., Halin, N., Green, A., & Dimberg, K. (2011). Open-plan office noise: Cognitive performance and restoration. *Journal of Environmental Psychology, 31*(4), 373–382.

Kar, G., & Hedge, A. (2020). Effects of a sit-stand-walk intervention on musculoskeletal discomfort, productivity, and perceived physical and mental fatigue, for computer-based work. *International Journal of Industrial Ergonomics, 78*, 102983.

Kar, G., & Hedge, A. (2021, January). Effect of workstation configuration on musculoskeletal discomfort, productivity, postural risks, and perceived fatigue in a sit-stand-walk intervention for computer-based work. *Applied Ergonomics, 90*, 103211.

McCulloch I. (2002). Health risks associated with prolonged standing. *Work, 19*, 201–205

Mellert, V., Baumann, I., Freese, N., & Weber, R. (2008). Impact of sound and vibration on health, travel comfort and performance of flight attendants and pilots. *Aerospace Science and Technology, 12*, 18–25.

Segal, M., & Hedge, A. (1999). *Space shuttle noise and active noise reduction effects on speech intelligibility.* Proceedings of the Human Factors and Ergonomics Society 43rd Annual Meeting, Vol. 1, pp. 41–45.

Sengupta, A. K., & Das, B. (2000). Maximum reach envelope for the seated and standing male and female for industrial workstation design. *Ergonomics, 43*, 1390–1404.

Vellinga, R. (2001). Research on sit/stand tables (in Dutch: Onderzoek Zit/Sta-tafels). Project-Inrichtingen.

Yates, J. W., & Karwowski, W. (1992). An electromyographic analysis of seated and standing lifting tasks. Ergonomics, 35(7), 889–898.

6 Do You Really Like Traveling?

6.1 ONLINE TICKET FOR PUBLIC TRANSPORT

Buying your tickets online sounds great. If you buy your ticket online when you go to another country, you will already have booked your travel at your destination. That is convenient. However, some public transport organizations also want you to book a time slot. That means that days before your arrival you must specify your moment of wanting to use the public transport. That's weird. A flight or train journey between two countries will never arrive exactly on time and if you do arrive on time there can be an unpredictably long queue at customs. Or your suitcase doesn't arrive when you are at the baggage carousel. It also always seems that all the suitcases of others come first. Many suitcases pass by and some look like your suitcase, but unfortunately, they are not your suitcase. At first, it is busy at the baggage carousel, but then after a while, you are standing there with just a few other people, who all think "what about my suitcase? Have they lost my suitcase"? You also often have no idea how far it is from clearing the airport to where you get your public transport and how long it will take. There goes your time slot for the public transport you booked. Now you do not only worry about your time slot, but also the suitcase arrival is an issue to worry about. Earlier in your journey on the train and in the airplane, you sometimes see the delay increasing and instead of being happy that you bought an online ticket you become more and more nervous. And you have all the time to think about it because the delay is increasing nicely. What a misery. With some bus or train companies, you can only buy a ticket online. If you did not book beforehand then you are standing at the bus entrance next to the bus driver and he looks at you like: "idiot, what are you doing here next to my steering wheel"? There is no possibility to buy anything on that bus. The same goes for the train. Your ticket that was valid for the previous journey has become worthless. Often there is no staff around to offer help, so where can you find a vending machine, or how can you change your ticket? The annoying thing is that it is not a moment of trouble. No, the entire journey before you get to the place where you booked the public transport you are in trouble. Will I still make it? Or are you too late anyway? Can something still be arranged? It is often difficult enough to find the exact location of your bus or train at an airport or station. Sometimes there are signs at the airport, but sometimes the bus or train is not on them and hard to find. And once you have found the place, at the bus or train station, you must figure out exactly where your connection departs. Then you are also late, and your time slot has expired. Your entire journey can be ruined because of this. During the flight, the flight attendant asks, is there anything else I can do for you? Then you think, yes of

DOI: 10.1201/9781003637035-6

FIGURE 6.1

course, put my suitcase first on the baggage belt on time, do not make the customs line too long and organize the connection at the airport. In the end the taxi, Grab or Uber, still seems the most comfortable and reliable, if the lines aren't too long and the waiting time isn't too great.

There is much literature on how to place the passenger journey central and not the profit of the individual companies, but in practice, this is hard to arrange. The number of publications about mobility as a service (MaaS) has increased rapidly in the past years (Maas, 2022). Additionally, there are new mobility services, like bicycle or electric scooter rentals and new train connections, which reflect innovation in the mobility market and have led to a diversified market landscape. But as the variety of options increases, it makes MaaS more complex and it is hard to combine all stages of a journey onto one electronic platform. While travel planning can now be done quite well, dealing with delays, making one overall booking and changing the booked travel segments is still an issue to be solved. Veneman et al. (2020) describe that:

> monopolization of certain travel service modalities has occurred, either because the operators are state owned and do not face competition, or because travel service modalities have been captured completely by powerful commercial parties (e.g., just one taxi company, just one bike sharing company).

This has the consequence that a traveler, will face various companies, each with their own electronic system. These systems now in practice are designed by humans. Now it is time that the designers, enterprises or governments place the passenger journey central. Of course, there is some progress in this area showing that it is possible to put the passenger central (e.g. Kefalidou et al., 2016), but there is still work to do to reduce the misery of the traveler.

6.2   THE MIDDLE SEAT ON THE PLANE

Oh dear, you think, "I'm sitting in the middle seat in the airplane". Luckily, there's no one sitting next to you. That's what you think when you get on, but then several people pass by, and you feel the threat that one of them will come and sit next to you. There are some really potentially annoying-looking passengers, too neat, too large, too tall, too smelly and some look like a chatterbox. You don't want any of them next to you. And yes, there they are: your neighbors for the flight. One is an incredibly large person who doesn't fit in the aircraft seat. The armrest has to be up to get this person into their seat and they need a seatbelt extender. What a nuisance. You know you will constantly feel pressure from a stranger against you. A strange body is pressing against you, you don't want that. Fortunately, your other neighbor in the window seat is a reader, but this one immediately confiscates both armrests. An overweight neighbor settles into the aisle seat and also immediately confiscates both armrests. Neither neighbor seems familiar with "armrest etiquette" – the middle seat should get both armrests (Elliot, 2020)! But you now have neither armrest. As a consequence, you have to sit with your elbows in front of your stomach. And stay like that the entire flight! Now another problem arises. How on earth are you supposed to fasten your seat belt. The two ends are nowhere to be seen. You can't move yourself. You can't look under your buttocks to see if you've sat on them. You have to find the belt somewhere by touch, because you can't see it. After a struggle, during which both neighbors angrily look at you as if you are committing the greatest sin, you find both ends of your seat belt. But on some planes, including the one you are on, there is only one belt and you have to push the end of your seatbelt into a buckle to lock it in place. But where is the buckle? Has it vanished between the seats? You look for the buckle opening again, but there is the overweight neighbor. How do you solve this? Just touching a stranger's buttocks is very inappropriate. There is a chance that you will be taken off the plane after involuntarily touching your neighbors' buttocks, led to the airport police and given a hefty fine. It will probably end up on social media and you will be portrayed as a hands-on person, a molester who can't keep their hands off other people's buttocks. You will lose your job. People will look at you on the street: "There goes that groper". You will soon be sitting alone in your room

FIGURE 6.2

moping and you won't be able to go outside. You are in a difficult situation. Maybe a solution is to not put your seat belt on. However, if you don't put on your seatbelt, the flight attendant might get angry and maybe you will have to get off the airplane for that reason. So you think you will take the step to ask your neighbor for a favor. You will ask it in such a way that the overweight person in question will not feel hurt, because if you are seen as being rude, you will still see later on social media what a terrible person you are. How do you do something like that politely? Many of us have always learned to keep things to ourselves and maybe not blame others for problems. But that's not the case now. Pretending to be a bit clumsy might also help. So the best solution is to politely ask: "The flight attendant will ask in a while if I have my seat belt on. Can you help me put the belt on"? And then hold the beginning of the belt in your hand, very helplessly. You might get the answer: "figure it out yourself", but you are lucky the answer will be: "I'll put it in". Phew, that's solved. Now you just need to hold your arms in front of your body for a few hours and then this misery will be over. Sometimes flying can be so awful.

These issues can be solved, of course. Special seats for wider persons and better seat spacing for accessibility and visibility. For the seat width design, it is important to know that the hip breath is not the widest part of the human body. The elbow width is the widest part of the human body (Molenbroek et al., 2017). Another thing is: do not place the seat belt on the seat, but to the side where passengers will take their seat if they cannot sit immediately, and when they sit, they need to see the safety belt. Another solution is "staggered seats" as is shown in the picture. Liu et al. (2021) showed that staggered seats have the advantages that you do not sit shoulder to shoulder and have your own place at the armrest, which resulted in more comfort, which is not a surprise. That is what we want. But this arrangement makes conversation with the person next to you more difficult, and if that is your spouse or a friend, you do not want this. Also, staggering seating in an airplane is not that easy as space is fixed and maybe even some seats will be lost in the front or back of the airplane and ingress/egress is more complex. Although the staggered seats in the picture have a seat pan which the front can rotate upwards and fold on top of the back part of the seat pan and then there is even more space for ingress and egress than in current seats. And staggered seating isn't needed for smaller planes with just two seats on each side.

6.3 OPENING THE SINK DRAIN

What a beautiful sink in this hotel room. It looks nice and shiny, clean and handy to rinse anything in. The hot water faucet is usually on the left. So, if you just turn it on, there will be warm water for washing. But it's taking a long time, a very long time and it still isn't warm. Too long. Is the warm water system working? Or what is the issue? Is it the correct faucet? After a while with no warm water, you decide that it is worth trying the other faucet. And now, warm water comes out of the right faucet. But there are no labels on the faucets. Usually, there is a red dot or cap and blue dot or cap indicating hot and cold water somewhere on the faucets. This helps because usually red is warm and blue is cold, but that's not the case here, no color cue is visible.

FIGURE 6.3

Well, after a lot of fiddling, there is warm water. Now close the sink drain and fill the sink with water, then rinse things. Afterward you'll want to let the dirty water run away. Then you think, just press the sink drain stopper, because that's often the case. Which is not handy by the way, because if there is very hot water in the sink you may scald your hands. Did they think of that in designing this plumbing? Or when you've just dried your hands, they do have to get wet again. But anyway, in this case pressing the top of the drain plug doesn't work. Oh wait, there's also sometimes a control behind the faucets. You can't easily see but you can just feel if there is something there that you press down or pull up. But no, there is nothing there either. What misery. How on earth does that sink empty? By now you might be getting a bit impatient. What a mess, who came up with something new again without thinking that a user should understand it. So, someone once came up with the thought "everyone drains the sink by pressing the sink drain or by pressing something behind the faucet, but I am going to design it differently, because that makes me happy". The fact that the users are completely confused by this, think they are idiots, or curse, or even worse, maybe break the sink, is something the designer did not have in mind. So, should the user scoop the sink empty with a cup? And where should that water go? Oh, maybe you can empty it in the shower or the toilet. But there must be a way to drain the sink. You may think to yourself "I am now going to lie under the sink and look up with my head to see if there is a connection to the sink drain system". Would all guests in this hotel do that? Aha, there is a cable from the sink drain to the back of the sink and that runs back to the floor and what do you see? There is a button on the floor that you can press with your foot. It is positioned on the floor right against the wall. So you crawl out from under the sink again and press the button with your foot and yes, the sink empties.

Norman (2006) has neatly described in his book *The Design of Everyday Things* that for pleasant use, there must be "consistency". That is to say that it is useful if a product works in a similar way to other similar products. That would certainly not have been a bad idea with this sink. He also writes that "visibility" is useful. It is good to have controls in sight. With this sink you had to look under the sink and then discover a floor button, which is not directly in the line of sight. And on top of that, why would anything think that pressing a floor button is linked to the sink drain and meant to let the water leave the sink. And if the user is in a wheelchair, how do they press the floor button – run over it? In this case user research, testing the product that is designed, would also have quickly revealed that this design simply does not work well (Baxter et al., 2015).

6.4 THE CONTROLS

Unclear signs can confuse us and make us feel miserable. On an airplane seat, the controls can be located on the seat arm where they cannot easily be seen. The pictures below show the left side of an aircraft armrest viewed from the right, and the picture on the right side shows a panel of elevator controls. At first sight, they seems OK, but when you want to use them, then it becomes a different story. First of all, the airplane seat buttons are against the passenger's thigh. That means the view is blocked. If the passenger didn't notice them before sitting down they will not be aware that there are buttons and controls in the armrest. Maybe they could feel the buttons with their thumb but they can't see what each button controls. If they randomly press a button, their backrest might recline and maybe that will annoy the person seated behind them. There are many tales of seat recline battles on airplanes! And if the person behind you is angry, which you cannot see, and you have no clue what that person is doing, then maybe they will push his knees very hard in your back or do something worse, like pouring a drink over you. The ability to recline your airplane seat is important, especially on long flights or with poorly padded seats because reclining the seat helps to distribute your weight, some on your thighs and some on your back, which reduces back stress and improves comfort. Not all planes have seats that recline, some have "pre-reclined" designs, but these are intended for shorter duration flights. So while the reclining button is nice for you, reclining your seat may not be good for the person behind you, especially if the seat pitch is tight or if they have a laptop on their tray table. Apart from the recline buttons, what about the invisible and hard-to-reach parts of the armrest controls, there may be a hole

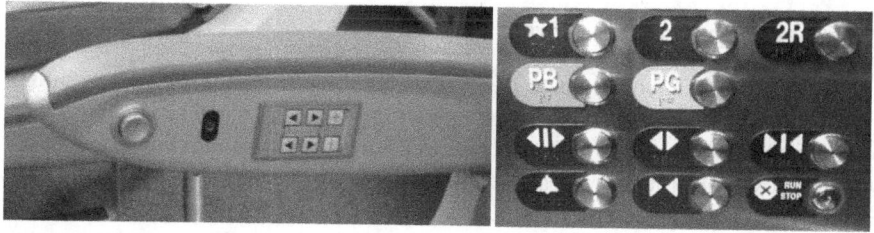

FIGURE 6.4

for a wired earphone jack and there may be plus and minus buttons for the volume, but what are the buttons with the arrows for? For choosing the channels of music or the videos playing on your screen, if your seat has a screen. But why are there four arrows? It is a mystery. Someone designed this, there should be some kind of logic, but for the user, it isn't obvious and does not immediately "ring a bell". It is always best to see what task the user wants to do and design controls, such as buttons based on the tasks in a way the user will understand it.

And what about the elevator controls in the right picture, they are also very unclear. If you want to go to the first floor that might be rather clear, but it isn't because it depends on where you are; in the USA "1" indicates the ground floor, whereas in the UK "G" is the ground floor and one is the first floor. And with the elevator button panel shown above for the second floor, there are two buttons, and unless you paid close attention to where you parked your car, you probably have no clue which one you should use. Presumably "2" is floor two (although in the UK this is actually the third floor) but does 2R indicate the right side or the rear door? Your car is in the parking garage, but it is also unclear which button you should use for the parking garage. Should it be PB or PG? And if you want to close the doors of the elevator, there are also two possibilities, and you have no idea which one you should choose. In this case, the designer might have had an idea on how to organize controls, but it is important for the designer to think of themselves as a user. The designer of the controls should simulate what the user has to do, like Kortum (2022) described. If the user wants to go to another floor then place labels on the buttons for that action, the user will immediately understand. In the elevator, you do not have much time to decide, else the elevator may start moving without you having pushed a button.

The principle of mapping the controls to the user's actions has been proven to work (Norman, 2006). It means that on a control panel the buttons are placed in such a way that they have a relationship with the environment. In this case, it would make sense to have the lower floor buttons lower on the panel and the higher floor buttons higher. Then still it should be one button for floor two and one for the parking, unless, as in this case, the elevator has two sets of doors that open either to the front or to the rear, hence the buttons with "R" labels – but there is no sign telling the user about this.

So, the misery of this user using the controls can be easily improved by checking what an end user would need, what this person can see, mapping the buttons, and limit the options to those that are relevant.

6.5   CHECK-IN AT THE HOTEL

After a long journey, you want to be in your hotel room. But what is happening now? There is a line at check-in. That is annoying, well then you have no choice but to stand in line. After some patience, it is your turn. And then the desk clerk says: "you are lucky your room is already available for early check-in. We have the extra service of early check in. If you pay $20 you can go in your room now", what is the extra service? You can come back in 20 minutes and then it will cost nothing extra to check-in. After some thinking, you decide not to pay the $20. You will just grab

FIGURE 6.5

a cup of coffee. After 20 minutes, you go back to the front desk, but now the line is even bigger. What a mess. So, you must go to the back of the line again but after 30 minutes waiting it will be your turn. Once at the desk, you must fill in all kinds of information that you had previously filled out on the internet. Who designs a system like this that duplicates effort? The hotel already has all the data from you but now you must fill it out again on paper, and presumably, someone will then have to transcribe that paper into the hotel's electronic system. Surely the digital information is much more convenient for the hotel. And the questions are so silly: when were you born? What does the hotel have to do with that information for the second time, as it was already provided when the electronic booking was made? And what is the date of arrival. This is a completely ridiculous question. They see you checking in now, right? You have already filled the date online when booking and the day of check-in is now. Does the hotel still have to ask what the date is today? What kind of people designed this system; do they assume that the hotel staff won't know what day it is and do all the people who check in have to fill in that date and is the hotel organization then convinced of what date it is? How much proof is needed to know what the date is? Shouldn't the software generate an automatic date and time stamp when you check in? Apart from potentially having some security implications, being asked to physically write down one's address, passport number, telephone number and email when it is digitally in the system seems unnecessary duplication and it slows the whole check in process for everyone. That's how you create a long queue at check-in. So, someone in the hotel has decided that you must pay extra if you arrive 20 minutes early, that you have to write down all the dates manually for the second time and that they ask nonsense questions like "what date is it today"? Yet these kinds of hotels also get good reviews on travel websites. So, why do we all just accept that someone in the hotel designs these kinds of ridiculous rules? Complaining to the desk clerk is pointless because they only obey orders and carry out what others have thought up.

After a long journey, you want to relax in your hotel room, but with the additional check-in frustration this is now not easy. No, you must wait and have to be endlessly patient. A similar situation can occur when renting a car. Also, at car rentals, there can be large queues and they ask about who you are and want to see your driving license, yet you also have to write down again what you have previously digitally entered into their software system. The counter staff in hotels, at car rentals, and the like are behind computers and these computers have software that is apparently often slow, because they have to spend a long time searching and entering data. Also, like at hotel check-in, at the car rental office, there are sometimes long lines, because of arcane administrative rules and slow software designed by people that do not have the end user and the client in their mind.

There are many papers that show a relationship between waiting time and customer satisfaction (e.g. Vinish et al., 2022; Liang, 2019; Pruyn & Smidts, 1998). Waiting negatively influences satisfaction and product evaluation. So, there is room for improvement, and it is possible to make customers more satisfied. However, in hotels, we probably forget the frustration after a while and our ratings on websites or social media are still ok. Probably because our memory is limited and studies show that after a while we remember the end situations and only some details of the total experience (Goossens, 2024), but usually we don't remember the beginning. Still, there is a need for owners of hotels, car rental companies and the like to pay attention to the check-in experience and reduce the user's misery.

6.6   STARTING THE CAR ENGINE

That is a nice rental car. You see it from a distance. Just push the key button on the fob to open the door. But what happens now? The trunk of the car opens as well. Well, that's solvable, you can just close it again. Then you can sit down in the driver's seat and start the car. But first, you need to adjust the driver's seat so you are comfortable. So you search for the seat controls because they are different for every car. Designers apparently like that variety. For each company and car model, designers seem to think they need to do it differently, because that's better and gives the car a unique style. But okay, the seat is now adjusted so you can just press the "start" button and go. But where is the start button? Usually, somewhere at the right of the steering wheel above the center console. Oh, this time it isn't there, you need to do some searching. Aha, there it is, hidden behind the steering wheel on the right side of the steering column. So, you must lean your head to the side to see it. Visibility of controls is very important, but apparently, not all designers have that logical principle in mind. And if you are left-handed, the start button is not in an optimal location. But now you've found it, you can start the car. Simply press the button. But when you press the start button, nothing happens. What now? Look up the manual? Oh, wait a minute. In some cars, you must also press the clutch. Oh, but this is an automatic – no clutch. So you pause to think. Maybe you need to press the brake and then the start button. That doesn't work either. Boy, what a hassle. There you are, in a multistory car park in a foreign country, feeling helpless. It's good that you're alone, imagine if your family were there, impatient kids, and you looked incompetent. Then you'd be marked for life with a reputation as a clumsy wimp – can't even start a car! Everyone is sitting in the car waiting for you to get the car going. You are in a warm country in a warm car and without it starting, the air conditioning doesn't work, so now you are sweating and starting to feel very uncomfortable. You've been working on it for a while now before you can even drive your rental car. Of course, you know your own

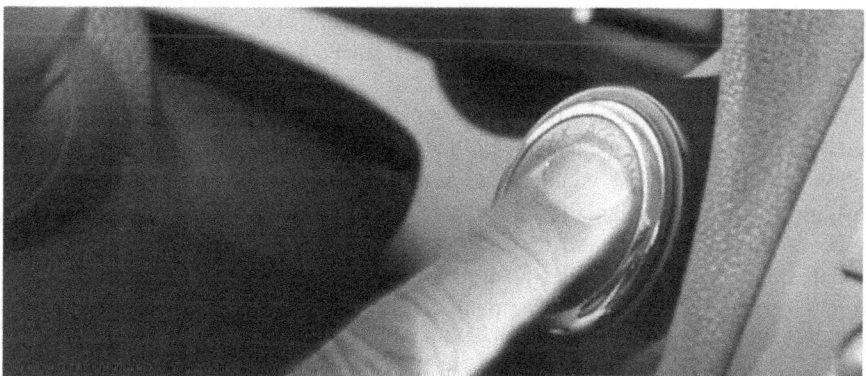

FIGURE 6.6

car well, but to get to know all the cars you could drive, you probably would need is a few years of study, and designs can change with every new year and new model. Is there a course on "dealing with diversity in cars"? Of course, there are courses on "dealing with diversity in people", but apparently that's not really relevant when we are dealing with the array of technological designs. So now you either have to ask another motorist who has rented a similar model or go back to the car rental desk or find car rental staff and ask them how to start the car. Either way, you will feel embarrassed by your incompetence. So you persist and suspect that eventually, you can figure it out for yourself by trial and error. So far there's a lot of error and you feel like a klutz, while it's actually just a nonintuitive design that is the problem. You think more and reluctantly decide you'll just use the manual if you can find it, after all, it is a rental car and the manual may not be missing, but it is missing. But then you realize, you can resort to your trusty phone and check the internet. And yes, there appears to be a YouTube video on how to start this car. You were remarkably close to success. Your only error was that you must put the automatic transmission from the parking position (P) to the driving position (D). You had tried that action, because have driven an automatic transmission before, but with this car, you can only go from P to D if you keep the brake pedal pressed, that is a special "safety feature" with this car. So now you press the brake pedal, put the transmission in position D, and then use your finger to press the start button. And yes, the car starts. Yipee. The screen comes to life and in the navigation system, you can enter the address, where you want to go. You have to enter your destination at the bottom of the display but read what is happening on top of the display. So, your eyes keep moving back and forth, which is not easy, but you manage. The acoustic confirmation of the letters does not help much in that respect either. Your destination is an airport and not a street. But there isn't an "Airport" option in this navigation system. The system does not recognize it if you input "airport". Wait, then you think "I am in Germany, let's try 'Flughafen'". That also leads to nothing. The city of the airport does work, and if you then type in 'Flughafen' it becomes 'Flughafenstrasse', well then I will select that and start the navigation system. That will probably be near to the airport, and you hope you can probably find the airport via the road signs when you are closer to the airport. Now at last you are ready. You can start driving. But the experience has been a bit frustrating, and taken quite a long time to figure out.

An important point here is that there is no design consistency in either to controls and operations required to start the car and the operations needed to start and navigation system and choose you destination. Consistency and visibility are important design principles (Norman, 2006). But user research on these activities would also be useful here as well (Baxter et al., 2015), not just among users who have this car model already, but with people who don't know this product at all. Often in car companies, studies are done using company employees. That is easier as you may not need NDAs (non-disclosure agreements) and also you can better protect confidential design development information, but it has the disadvantage that you may not find the problems described here.

6.7   THE SMART ENVIRONMENT

Sitting on the toilet at work, you are suddenly plunged into the dark, and you realize you have to wave with your hands to get the light on again. It is good that no one sees the silly movements you make while sitting with your pants down or skirt pulled up. When people see you like this, you will not convince your colleagues in the boardroom or enhance your reputation in front of a classroom. I think many would think you look foolish.

Sitting there in the restroom (toilet) is already a potentially embarrassing situation and now you also have to wave your hands a bit stupidly, it is a good job that people you know are not staring at you. When this "misery" is over, you will want to wash your hands. You probably have to touch the handle of the toilet door, which probably is not hygienic and you think after opening the lock, which you have to touch as

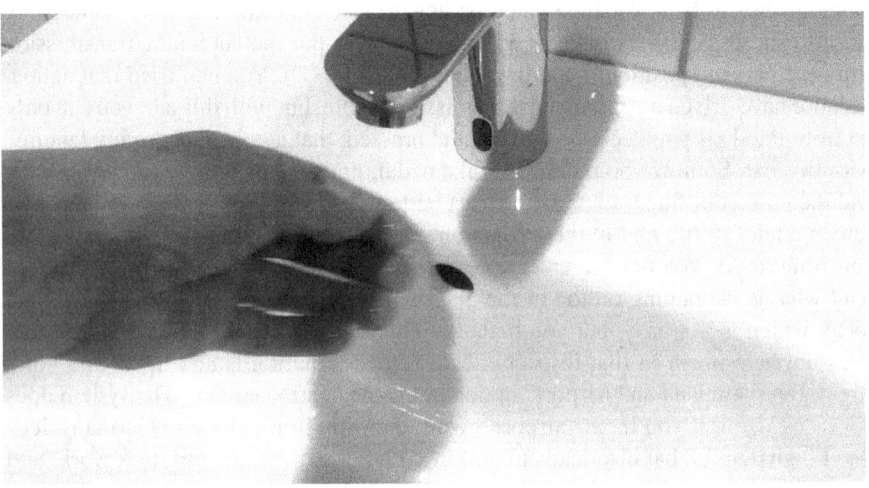

FIGURE 6.7

well (Door handle bacteria - Wikipedia). So the next step is to wash your hands. You want to do that with soap and water of course, but systems do not always understand that. In busy places, the soap dispenser is often empty and sometimes it is still quite a feat to get the water out of the tap. Where is the sensor. First try under the tap. No, that does not work. Maybe to the side of the tap? No, that does not work either. Hold your hands above the tap? No, that would be very illogical. By the way, how do you regulate the temperature of the water? Could there be a sensor that thinks: "you are that type of person who likes this temperature of water". That should be possible, of course with the current AI. But you have not come across it yet. For the time being, you are still fumbling around with "smart" systems. And you still have no water. The sensor also seems to be in a different place for every tap and the time it takes between the sensor activation and the water to come out differs as well. While there is enough literature that states that consistency (Norman, 2006) is an important design principle as well as immediate feedback. Finally, you accidentally touch the tap and the water comes out, you don't know what you did, and the water stops quickly. Again, you try all kinds of possibilities and, after another accidental touch of the tap a bit of water is there. So quickly you wash your hands. You have the feeling that many people around you in the washroom are staring at you and are puzzled by the crazy movements you have been making to get the soap, water and dry your hands.

The use of a smart water tap might be more hygienic if you did not have to touch the knobs of the tap or the tap itself. But the lack of any signs or instructions introduces many problems: first of all, users must experiment to find the place where they get the water running; secondly, users cannot control the temperature; thirdly, if nothing works, users cannot determine whether or not the tap is broken. The more electronics are in the system, the more possibilities that something can break, such as sensor malfunction, "dead" batteries, valves not working anymore, the cables that are broken, etc. The hygienic advantage of not touching a tap has evaporated! By the way, if the bathroom has doorknobs, these need to be touched as well. So, automatically opening doors should be added or corridors that also can prevent people outside the bathroom from looking in.

Nevertheless, accurate positioning of the sensor and some instructions certainly need attention. But there is more. Regardless of the positioning, the design must follow universal design principles like "affordance" (the object characteristics that suggest how it can be used), and respond within 500 ms to be perceived as instantaneous and functional for an end user (Hansen, 2016). Technologically, we are capable of doing this, but making many kinds of systems convenient for end users still needs a lot of ergonomic design attention (Norman, 2013).

6.8 TRANSPORTATION DELAY

You might have an app on your phone for the local train company, and you can see whether the trains are on time. This is a good service. Let's say you want to take the 8:00 am train, you know you have a 20 minutes bicycle ride to the train station and your first meeting is at 9:15. You think you have plenty of time, and you hate

FIGURE 6.8

people that come too late to meetings. You could travel by car, but morning traffic is unpredictable, so the train is more convenient and almost guaranteed to be on time. To avoid morning traffic congestion, you decide to cycle to the railway station, only a 20 minutes ride, then you park your bicycle, lock it (which is certainly necessary as you will need the bike on the way back as well) and wait at the platform. But what has happened? The train is now delayed. But 20 minutes ago, when you left the house it was on time. How is this possible? The signs at the platform just say "delay". Any information on the duration of the delay is missing, which makes you feel increasingly insecure as time passes. You start to feel stressed. But then there is new information on the station signs. Twenty minutes delay. Phew, only 20 minutes so the train should arrive with 20 minutes to spare before your meeting which is only a few minutes from your destination station. So, you should still be on time. You check your phone for confirmation. The phone app now just says "delay". Which information should you believe? Oh no, after another ten minutes the station information changes to a 25 minutes delay. It is getting tighter, but if everything goes like clockwork, you will still be on time. And then five minutes later there is misery, now the station sign says the train is canceled! How on earth is that possible? The train was running as it was delayed and now the train is gone. Has the train completely disappeared? Did someone with a large magnet take the train from the track? Or just like in some cartoons did it suddenly disappear? Is there a secret switch and the train with all its passengers is gone? Has the train been abducted by extra-terrestrials and transported into an alien world, and no one will ever see them again?

It is a mystery, how a train can suddenly disappear in a few minutes. A complete train is quite an object, but according to the station sign that large entity is gone. At some earlier time, it was on the track, of course, later than planned according to the signs, and it was on its way to your platform but now the complete train is gone. I don't think many of my fellow passengers waiting on the same platform will understand this either. If you had known this the moment you left your house, you could have taken your car. But every solution you can now think of leads to misery, you will simply be too late for the meeting. And the station displays offer no alternatives for passengers.

In this case, the train time and progress information was not handled well and someone within the train company decided or programed the information systems in such a way that this vanishing trick could happen. There is also no communication between the app and the system at the platform. It is hard to test each situation with end users, but a cognitive walkthrough can be done by the designers or fellow designers as well and might have helped here. The cognitive walkthrough with different scenarios of delays and how to inform passengers might have helped here (Lewis, 1982). Anyhow, going through different scenarios is relevant and might prevent these situations. Consistency can help as well (Norman, 2006). The app on the phone should have the same information as the signs on the platform. These systems should use the same source of information. This is not only the case with trains, of course. Also, for airplanes, the app and the information at the airport can differ. This again creates misery. Sometimes you can see on flight radar that the airplane will come in very late and then still at the gate the information display says it is "on time". It is interesting that we have several systems that do not communicate which each other. But people designed each system, they just never communicated their designs between the systems. Often people working in different companies have to follow their own systems design guidelines and rules, and often these don't easily translate between systems. So for the end-user, the passenger in this case, journey information is confusing, the lack of suggested alternatives is distressing, and the system failure simply leads to misery.

6.9 STUCK IN THE AIRCRAFT SEAT

This is terrible, an 11-hour flight and no space to move. How do you survive this? You will be stiff, have back pain, run the risk of a deep vein thrombosis (Adi et al., 2004; Şabanoğlu, 2021), and feel uncomfortable and miserable. And feeling uncomfortable is not short term, no, it is long lasting and gets worse over time. After sitting in the seat for a little while, you begin to feel a bit of discomfort and you know it will continue and worsen during 11 hours of the flight. And you cannot be the only person feeling this. You are surprised that after four hours, not everyone is crying in the airplane. Well, of course, not everyone. Those in first class, business class, premium economy (if there is one) and the very small people and children can handle the cramped space. Will all other passengers tolerate and survive this misery? Did we, economy class passengers, realize we were paying to be tortured for 11 hours! Did we realize we were risking our health because of deep vein thrombosis risks? To enjoy the relative spaciousness of first or business seats or even premium economy is an enormous price difference, often many thousands of dollars. But how can you

FIGURE 6.9

know how cramped your seat space will be before you board the airplane? Have you ever bought new jeans and tried to put them on, but they are too tight. The salesman may try to reassure you that they will stretch in the course of time and if you wear them they will conform to the shape of your lower body. You may really like the jeans but they are the wrong size for your body and there aren't other sizes so you might try lying on the floor, breathing your stomach in and pulling the jeans up over on your hips. It is a bit extreme but you try it, and you get the jeans on. Now you can hardly breath and stand up. But the salesman is convincing and says, you should buy them because they are the perfect fit and they will stretch and loosen over time as you wear them. So you buy the jeans. But is it the same with the aircraft seat. In fact, yes, it does not fit you, but you have bought it. So you have to tolerate the discomfort and you hope that you can be distracted from your misery by watching a good movie. Well, this could be the case, but you might discuss that in reality such distraction might be unethical. Discomfort is a warning from your body that something is going wrong, and in this case, it could lead to musculoskeletal injuries and if we can't move around for 11 hours it could lead to deep vein thrombosis which could be fatal.

The profit of airlines is not that large (Anjani et al., 2021) so they cannot give everyone a large seat. Of course, if you want more space you will have to pay more. However, you might not be able to afford this because such seats are several times the price you can pay so there may be other underused solutions may help. Bouwens et al. (2018) developed a system to let people exercise in a pleasant way in a cramped aircraft seat. After studying different movements that were intensive enough and not disturbing the neighbors, a new gaming system was developed, where the up and down movement of the legs was used for controlling the game. The system is light-weight, only a few grams and can be easily connected to the in-flight entertainment (IFE) system. This system uses variation in leg pressure on the bottom cushion to control a balance game. Passenger's leg movement is registered by lightweight fabric pressure sensors in the seat pan. By lifting one of the legs or extending one of the legs forward, participants can control the video game where a ball is rolled left, right, forward or backward in order to collect small blocks that are placed on a certain path. When all the blocks are collected without hitting the walls, the participant continues with the next level. The effect of playing this game on movement and comfort was tested with 12 participants. The participants were sitting twice in an aircraft seat for 3.5 hours. In one case, first time six participants played the movement game for 5 minutes every 30 minutes and in the other case they were just sitting (the order was varied). The other six participants had an opposite treatment order. The effects were clear. The number of movements was 3.5 times more in the gaming condition and the comfort increased significantly. An example of a comment of a participant: "It also has some fun. It is more entertaining. It makes your flight a little bit less serious. I don't know how to explain, a bit more relaxed, a bit more comfy". This is an example of one of the underused options to make flying a bit more attractive.

I think we all know that variation of posture is very important and there is enough literature which proves this (Vink, 2023). Perhaps, it is time to be a bit more creative and look for solutions to prevent misery and, of course, maybe we should fly less or pay more for the flights that we take and have some more movement space. Less flying with more space is better for the environment and your body.

6.10 CONSISTENCY

That is handy, at this train station you can only get in and out if you go through an electronic gate. You can't make a mistake, and you never forget to check out. You don't have to think about that. It will now become automatic, and you will check out unconsciously and it can't go wrong. Wonderful, you don't have to think about it. This seems like the ideal situation. But then the misery starts. At certain train stations, you can just get off the train and must look for a "check out post", a designated area where passengers must scan or validate their train tickets upon reaching their destination, essentially "checking out" of their journey and signifying the end of their travel on that particular train ticket. But not all stations have a checkout post, and this time you don't think about it, and you just walk out of the train station. But you must pay for this mistake. You get a fine. You won't get a criminal record yet, but it feels as if you are treated like a criminal. You will have to pay extra money because you have made a big mistake, namely you arrived at the destination station for which you bought your ticket, but you did not check out. They blame you for not checking out. Systems at other train stations just let you walk through an exit gate out of the station. But at this station you only see the problem later, you don't get a warning at the checkout post station, but you see the fine charged on your account, and now you can't do anything about it anymore. Most people then think "how stupid I was". But that is not the case. The station exit system design is inconsistent. If you are used to gates that alert you that to have to check out, you will develop a behavioral pattern, which becomes unconscious. In the Netherlands, there is no gate at the largest airport, and you really have to think about not forgetting to check out. Arriving at an

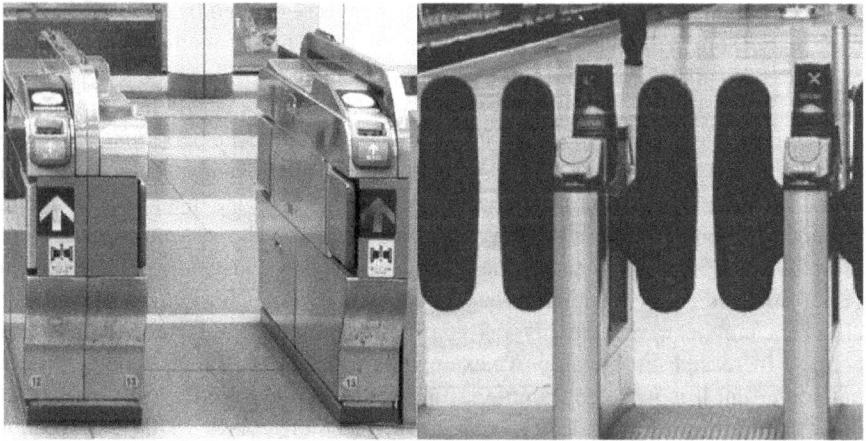

FIGURE 6.10

airport means thinking about your passport, ticket, looking for the check-in desk or gate. Some people are nervous about their flight. That requires your attention. This means that there is not much space left in your limited attention system. A person's average attention span is 8.25 seconds while for a goldfish it is nine seconds (Samba Recovery, 2024). Also paying attention to a checkout gate is a challenge because the design of the gate is not always obvious. At some gates, you must hold your ticket in front of a reader and the door that opens is just besides this reader, while for others you must insert your ticket into a slot. The exit gate ensures that you as a traveler will stop for a moment (Groot Obbink, 2016). In some countries, gates have the door 1 or 2 m further away. The advantage is that you as a traveler do not stop, and you do not have to wait for the gate to open. You can keep walking and that ensures that there are shorter queues when it is terribly busy. Another design aspect is the opening. By keeping the door of the gate open as standard, people also walk through faster. It only closes when someone threatens to enter without scanning or with the wrong scan. This also improves the flow, which means there are fewer queues when it is busy (Groot Obbink, 2016). But there are no gates at places with checkout posts and there are no warning signs, so it is easy to forget what you have to do to avoid a fine.

Of course, the importance of consistency in design among checkout systems also applies to many everyday things that a person uses. At the water tap, the idea is to make something that water comes out of in a controlled flow rate. Sometimes a rotary knob is used and this is usually reasonably easy to manage, though you have to remember the direction of turning the tap – clockwise for increase. Sometimes there is a handle that has to be rotated to the right, or sometimes moved forward or backward, and now we have the opportunity for confusion when people initially move it in the wrong direction. It becomes even more difficult if it is a mixer tap, mixing hot and cold water. Then you have to figure out whether moving to the right makes the mixing hot or cold. Or maybe it requires pulling toward you and then the other direction controls the volume of the water jet. It is a bit of a puzzle, but you usually figure it out. Sometimes with burned hands, but only a nitpicker would pay attention to that.

We are also used to a hinge when opening the door. So, usually the door turns inward for you to enter, but sometimes the door pulls outwards and sometimes it is a sliding door. If you make an initial mistake, it is not your clumsiness, but the inconsistency in design (Norman, 2006). We also know that there is inconsistency in power outlets when we travel between countries and also changes in voltage, so even with the correct adaptor plug you might plug a US 110V electric razer into a UK 220V outlet, and if the device doesn't have a built-in voltage regulator, you will fry your device! It's not just the plug but it is also the voltage converter feature that is needed. Difference in the design of products and technologies is not just inconvenient but it can also be dangerous leading to accidents. Design standards are useful and there are ISO standards that apply worldwide, but these standards don't cover every possible product or situation. So it is important that designers look at how the new design compares with what is already there and try to be consistent in designing the control and display interfaces possible to support end users and prevent misery.

6.11   THE SLEEPER TRAIN

How wonderful to get on the train at 10:00 pm in the evening, fall asleep, and arrive at your destination the next morning at 8:00 am, well rested. Isn't it wonderful, sleeping and traveling all night. You are moved in the time and space all while you are sleeping, and of course, the train is better than the plane in terms of sustainability, and there is no problem with turbulence on the train. What a good choice. When you board the train you are happy, because this is really a very smart choice you think. You have a reserved space in the train, which is nice as well. You just step on the train. Then you look for your sleeping place and you see that it is now still a seat, which must be converted into a bed for sleeping. The other passengers say that they want to sit for a while at first. And then seats can only be changed into bunk beds on top of each other if the seats disappear, the other passengers all need to agree that their seat is converted at the same time, so either all can sit or all can sleep. This is a disappointment. Now you are dependent on the desires of others. This is the time to convince the others that it is time to sleep on the sleep train. You are feeling tired but the other passengers look perky, so you may have to persuade the others to go to bed now. How can you do this? Maybe you can tell them that it is very unhealthy when you do not have enough sleep. You will make the others responsible for your bad behavior the next day when you did not have enough sleep. You tell them that you have to repair a nuclear power plant, and the work asks for concentration. If you do it wrong, a disaster will happen. Or you tell that them that these sleeping companions (they are not yet sleeping companions, but they will be soon) do not want to be responsible for the fact that you commit all kinds of terrible crimes the next day due to a lack of sleep or you tell them that you are a surgeon and that you have very difficult surgery tasks the next day and that if these kinds of activities go wrong, many patients will die. You tell the fellow compartment mates that enough sleep is essential, and they should really immediately change the seats into beds. But inevitably

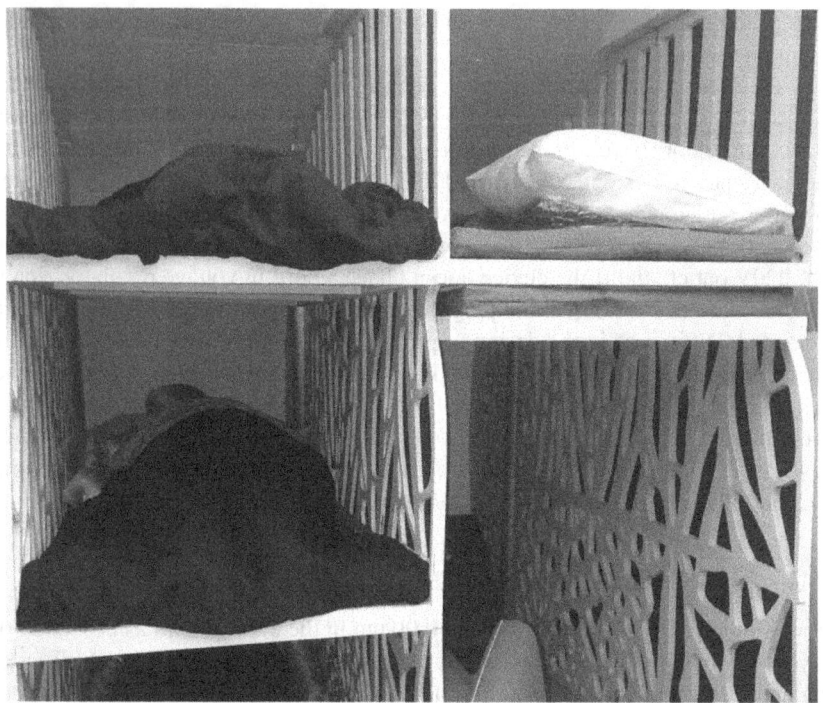

FIGURE 6.11

this speech does not work, especially, if some of the other passengers are young and lively. So, this lack of privacy and inability to sleep when you want is disappointing, and some of the other passengers are hard to convince. Their behavior is out of your control. Additionally, your luggage is open in the luggage rack. So, everyone has access to your luggage. That means you can't sleep deeply because you constantly have to check this to make sure it hasn't been tampered with or stolen. Placing your luggage in your bed, might be an idea, but the bed is already very small, and you have to wait for others to want to sleep. So, that is not really an option.

Fortunately, after an hour of traveling, the rest of the people in your train compartment also want to sleep. Then you all convert the interior from seats to beds and now you can all get into the bed. You have clean sheets. That's at least something. But you also have the upper bed, so climbing into it is a bit of a effort. But it works. Oh dear, the bed isn't long enough. So, you have to sleep with your legs pulled up. Oh dear, it's not wide enough either. After some struggling, you find an acceptable sleeping position. There are studies which show good bed sizes for traveling, but the designer of this train bed still have to read these studies (Vink et al., 2025).

After some time, the train stops. What a hassle. People get in and open the curtain. So, you wake up. It's annoying anyway that people can just enter the train at the station, because they can just get into my luggage. Hey, now you have to wake up at every station to keep an eye on your luggage. That is annoying. Then the train starts moving again. All this stopping and starting is annoying. We also stop when there's no station, what's that about? Sometimes we just stand still for ten minutes and also when the train is traveling the noise sometimes is really loud. Vibrations are also sometimes annoying. And accelerating and braking feels unpleasant as well. Then after a while, there's a loud knock on the door. Customs comes in the middle of the night to disrupt things. Why is that in the middle of the night necessary? Maybe they want to irritate us? Do they expect everyone to undress and so they can check if there are any drugs or other stuff in all accessible parts of the body. Do the customs officers put on gloves and cavity search some parts of the body to make the person really miserable. Fortunately, the custom visit is over in the end. What is also a bit annoying is that I suspect that some of the other passengers have been flatulent in the night. No one has showered or bathed or washed. So, the smell in the compartment now is not very pleasant either. Every time we stop, the air conditioning is turned off and it gets warm in the compartment. Making the smell a bit stronger. Couldn't that be done differently? Sometimes it is possible to sleep for a while, but it is not really comfortable. This experience is also confirmed by research. Vledder et al. (2023) found that "speed, temperature, humidity, noise, and the seat" all influence sleep comfort during a sleeper train trip. Abrupt changes in some factors, e.g., jerk or distinctive sounds were recorded as well. Especially, train acceleration and loud sound disrupted sleep quality. Out (2024) also describes that "People want a sense of security. This is not about the safety of the means of transport but about the feeling of security in the interior and knowing that your luggage is safe". There are of course conceivable solutions. One of the problems is that the night train has to run on a route where cargo traffic also runs at a lower speed, and getting on and off at different stations is difficult to avoid. So, logistically there is some work to do, but humans design the systems. So, it should be possible to improve it. But, of course, other solutions are possible as well, such as solutions in the area of soundproofing, good quality beds, storage places for luggage with a lock, better air conditioning, better washing facilities and good planning of the train journey so that you don't have to keep stopping all the time. Out (2024) has also described nice solutions, in which even ideas of business class seats from airplanes have been taken over and the idea that you combine the day and night train, so that you can sit comfortably during the day and sleep comfortably at night.

6.12   HAND LUGGAGE

Hand luggage is allowed in the airplane. That is great. Some airlines let you pay for the hand luggage and then you assume that there should be enough space for your hand luggage as you have paid for it. But why is it almost always a hassle with your hand luggage? Most of the time you can't put it away in the luggage bin as there is no space. And often, there is no overhead bin space close to your seat. Then you have to place it far away from your seat in the airplane. Then when you deplane, you have to go against the flow of passengers before you can get to your luggage or wait until all passengers have left the airplane until you can have access. And if you have a tight connecting flight, you now might miss it! Why does every traveler take so much

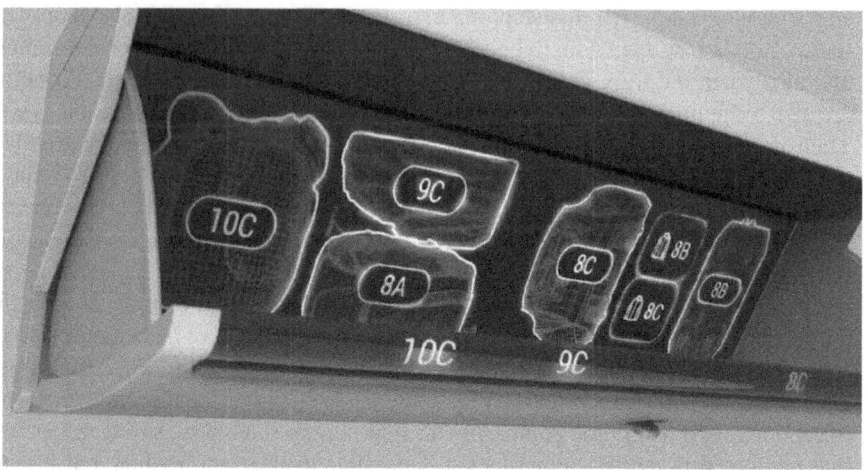

FIGURE 6.12

with luggage them? Maybe it's because in 2023 more than 2.8 million checked bags were lost by airlines in the USA alone (Hicks, 2024) and worldwide this was more than 26 million in 2022 (Parker, 2023)! But there are limits on what is allowed as carry-on luggage. And in the US, the check-in staff check the size and amount of your luggage. Typically airlines will announce at check-in that overhead space in the airplane is limited, that luggage size will be checked upon boarding and that large objects you still have must be gate checked at check-in. But still people have lots of "stuff". Many people still put coats, the stuff they bought at the airport, small and large backpacks in the overhead bin and who knows what else in there. Some US airlines now ask that small backpacks and bags be placed under the seat in front of you, but if you are in the first row or emergency exit row you have to put everything overhead. To make matters worse. In some aircraft, some overhead bins are reserved for safety equipment and cannot be used for carry-on luggage. So now there are more seats and people than overhead storage space! After the last passenger has boarded the flight, attendants often have to lug around the last suitcases to get rid of these and some of these end up in the hold, so then you can have extremely angry passengers who now have to collect their bags from the baggage carousel upon deplaning. Some passengers worry that their hand luggage is going to be handled roughly and may be damaged or contents may be stolen, especially if luggage has valuables or medications in it. And upon deplaning they have to wait for their luggage at the belt at the arrival airport. Oh, and if they have to gate check their luggage, they have to take out their medicine and laptop. "Take out medicine" this may be a bad sign that means my hand luggage will get lost. If you are sitting by the window, you may have forgotten to first take out things like your laptop and headphones before putting your bag in the overhead bin; so if now you want to work on your laptop or get your headphones out of your luggage, it is also a hassle for you and also the middle and aisle seat passengers in your row. They either have to exit their seats or you have to squeeze past everyone and once you have what you want you have to go in again and pass everyone to get to your seat. What is also so annoying is that while boarding, some people think they are alone on this planet. They calmly take all kinds of items out of their baggage while standing in the jetway aisle and blocking the path for others. They take out, for instance, a magazine, a laptop, a tablet, a cord to connect to the charger, their headsets, etc. Many passengers have to wait patiently behind this slow aisle occupant. This persons also takes something to drink out of their bag to go in the overhead bin, and oh yes, check if there is anything tasty in the bag for to eat. Then they put all these things on their seat so they can't easily sit down. And then they think for a while. "Well, I'm not going to read a magazine, so I'll calmly put it back in the bag". Still, I'd better check my luggage to see if there is anything else, I will need in the seat. Ah yes, my inflatable neck pillow that I will have to blow up! Finally, you hear the annoying person say "No, I think I have everything now". What a relief. The aisle is free again and now there is some movement making it possible to get to your seat. But then there is someone else who is sitting in the wrong seat. They are sitting in seat 18a, but this person's boarding pass says 17a not 18a. Unfortunately, all passengers sitting next to this person have to go back into the aisle and you have to back up so these people can pass the visual alignment of seat labels overhead with actual seat rows is often very poor in aircraft, so it is always a good idea to start counting the rows as you board the plane. Although some airlines have different row number designs and for sometimes row 13 is not even mentioned, because of the superstition that 13 is unlucky. Also, the numbers and letters often are small and sometimes hidden behind the door of the overhead bin, hard to see especially, if you are tall. So, there is still some room for improvement here. But boarding can of course become much more efficient and faster by better designs. Experiments have been done where people scan their suitcases at home and their boarding pass seat assignments, and an algorithm calculates how all the suitcases fit best (Coppens et al., 2018; Hiemstra-van Mastrigt et al., 2019; Vendel et al., 2019). An unexpected benefit of these experiment is that boarding also went faster. Not only because the luggage fits the bin space but also because the seats were found faster, because the hand luggage was close to the seat and people first put the hand luggage in the bin and then looked for their seat.

Unfortunately, most people first look for their seat and then they try to find a place for their hand luggage in the overhead bins.

Most airlines now have overhead baggage size checks at the check in desk. The advantage is that in an early stage it is clear (before the flight) that the luggage is going to be too big for the overhead bin space. Additionally, those travelers that check in last are told before they board that their luggage will be placed in the hold. This also saves irritation and hassle at the gate.

Some travel companies have a bag measurement app so you'll know what fits before your flight. Airlines usually have a bag size frame that you insert your bag into to test that it is appropriately sized for overhead storage.

These days some planes have taller overhead bins and passengers are instructed to place their bags side on rather than flat so that more bags can be accommodated. People still have to get used to this design, because now your suitcase has to go in on its side and not everyone does that. Changing behavior is difficult. But there are already many possibilities to improve this boarding process and the process of stowing hand luggage. Experiments have also been done with airplane seats that were narrower when boarding and disembarking. Then there is a wider aisle. And after boarding the seats became wider (Akkerman, 2016). People experienced that as pleasant, and boarding went significantly faster. Not only technical solutions and plans and working with an algorithm, but the behavior of people is also important in this case. Disciplined and trained travelers board much faster (Akkerman, 2016). And in the winter, it also takes a little longer because of taking off winter clothing (Spaargaren, 2018).

REFERENCES

Adi, Y., Bayliss, S., Rouse, A., & Taylor, R. S. (2004, May 19). The association between air travel and deep vein thrombosis: Systematic review & meta-analysis. BMC Cardiovascular Disorders, 4, 7. doi:10.1186/1471-2261-4-7

Akkerman, S. (2016). *Improving boarding efficiency and experience* (MSc thesis), TU-Delft.

Anjani, S., Song, Y., & Vink, P. (2021). Designing a floor plan using aircraft seat comfort knowledge by aircraft interior experts. Work, 68, S1.

Baxter, K., Courage, C., & Caine, K. (2015). *Understanding your users: A practical guide to user research methods.* Morgan Kaufmann.

Bouwens, J. M., Fasulo, L., Hiemstra-van Mastrigt, S., Schultheis, U. W., Naddeo, A., & Vink, P. (2018). Effect of in-seat exercising on comfort perception of airplane passengers. *Applied Ergonomics, 73,* 7–12.

Coppens, J., Dangal, S., Vendel, M., Anjani, S., Akkerman, S., Hiemstra-van Mastrigt, S., & Vink, P. (2018). Improving airplane boarding time: A review, a field study and an experiment with a new way of hand luggage stowing. *International Journal of Aviation, Aeronautics, and Aerospace, 5*(2), 7.

Elliott, C. (2020, September 18). As airlines begin selling middle seats again, it's time to remember nobody owns the armrests, *USA Today.* https://www.usatoday.com/story/travel/advice/2020/09/18/airplane-etiquette-who-arms-the-armrests/3478678001/

Goossens, R. H. M. (2024). *Application of the peak-end rule to seat discomfort.* Presentation at IEA2024, Seoul.

Groot-Obbink, L. (2016). Opening the closed: Design of a user-centered, closed payment border for public transport (MSc report), TU-Delft.

Hansen, S. B. (2016). *Exploring practical implementation of touchless access control using iBeacons in Norwegian hospitals.* Master's thesis NTNU, Trondheim. Norway

Hicks, J. P. (2024). *See which airlines mishandled the most luggage in 2023.* https://www.mlive.com/public-interest/2024/04/see-which-airlines-mishandled-the-most-luggage-in-2023.html

Hiemstra-Van Mastrigt, S., Ottens, R., & Vink, P. (2019). Identifying bottlenecks and designing ideas and solutions for improving aircraft passengers' experience during boarding and disembarking. *Applied Ergonomics, 77,* 16–21.

Kefalidou, G., D'Cruz, M., Sharples, S., Lille, C., Frangakis, N., Ottens, R., Grosmann, R., Marcelino, R., Lütjens, K., Löwa, S., Shaw, E., Nardini, A., Santema, S. C. (2016). *Passengers' requirements for developing a passenger-centred infrastructure to enhance travel experiences at airports.* Conference Proceedings of the Chartered Institute of Ergonomics and Human Factors. Daventry, pp. 1–6.

Kortum, P. (2022). Where's my jetpack? waiting for the revolution in statistical analysis software interfaces, but going in the wrong direction. *Interactions, 29*(5), 68–71.

Lewis, C. (1982). *Using the "thinking-aloud" method in cognitive interface design* (IBM Research Rep. No. RC 9265 [#40713]). IBM Thomas J. Watson Research Center.

Liang, C. C. (2019). Enjoyable queuing and waiting time. *Time and Society, 28*(2), 543–566.

Liu, Z., Rotte, T., Anjani, S., & Vink, P. (2021). Seat pitch and comfort of a staggered seat configuration. *Work, 68*(s1), S151–S159.

Maas, B. (2022). Literature review of mobility as a service. *Sustainability, 14*(14), 8962.

Molenbroek, J. F. M., Albin, T. J., & Vink, P. (2017). Thirty years of anthropometric changes relevant to the width and depth of transportation seating spaces, present and future. *Applied Ergonomics, 65,* 130–138.

Norman, D. (2006). *The design of everyday things: Revised and expanded edition.* Basic books.

Norman, D. (2013). *The design of everyday things: Revised and expanded edition.* Basic Books.

Out, A. (2024). *A day and night train interior design for improved passenger comfort and improved train usage* (MSc thesis), TU-Delft.

Parker, B. (2023, May 17). Lost luggage figures reach 10-year high as 26 million bags go missing at airports in 2022. *The Independent,* Wednesday. https://www.independent.co.uk/travel/news-and-advice/lost-luggage-airports-missing-bags-b2340454.html#:~:text=Lost%20luggage%20figures%20reach%2010%2Dyear%20high%20as,bags%20go%20missing%20at%20airports%20in%202022.

Pruyn, A., & Smidts, A. (1998). Effects of waiting on the satisfaction with the service: Beyond objective time measures. International Journal of Research in Marketing, 15(4), 321–334.

Şabanoğlu, C. (2021). The secret enemy during a flight: Economy class syndrome. Anatolian Journal of Cardiology, 25(Suppl. 1), S13–S17. doi:10.5152/AnatolJCardiol.2021.S106

Samba Recovery. (2024). Average human attention span statistics & facts [2024]. Samba Recovery. https://www.sambarecovery.com/rehab-blog/average-human-attention-span-statistics

Spaargaren, C. (2018). *Redesigning the deboarding experience* (MSc thesis), TU-Delft.

Veeneman, W. W., Van Kuijk, J. I., & Hiemstra-van Mastrigt, S. (2020). Dreaming of the travelers' experience in 2040: Exploring governance strategies and their consequences for personal mobility systems. In *Towards user-centric transport in Europe 2: Enablers of inclusive, seamless and sustainable mobility.* Delft University of technology*: Delft, the Netherlands* (pp. 225–239).

Vendel, M., Dangal, S., Coppens, J., Hiemstra-van Mastrigt, S., & Vink, P. (2019). Effects of a hand luggage guiding system on airplane boarding time and passenger experience. *International Journal of Aviation, Aeronautics, and Aerospace, 6*(3), 5.

Vinish, P., Pinto, P., & Hawaldar, I. T. (2022). Consequences of retail checkout crowding on perceived emotional discomfort and switching intentions. *International Journal of Innovative Research and Scientific Studies, 5*(2), 134–144.

Vink, P. (2023). *Seat comfort and design.* Pumbo.nl.

Vink et al., (2025). How do we sleep? Towards physical requirements for space and environment while travelling. *Applied Ergonomics, 122,* 104386.

Vledder, G., Yao, X., Song, W., & Vink, P. (2023, September). Explorative study for sleep conditions in sleeper trains. In *Comfort congress* (p. 151).

7 How Not to Fool a User

Everything we have described in the previous chapters has been organized, designed, engineered and made by humans. So why do they fail? It's simple; they ignored the ergonomic design principles that play a crucial role in creating user-friendly and efficient interfaces. Ergonomic design for usability focuses on creating products, environments and systems that are comfortable, efficient and safe for users.

We have the power to make better designs possible. However, it should be realized that it is often not the choice of one person, which can make changing more difficult. Sometimes it is not even the choice of one company. Dougali et al. (2014) showed that many companies are involved in organizing a travel journey, but they work in silos, which means that each company optimizes its own situation. Instead of focusing on the overall travel experience of the passenger, they form single points of contact with their customers, supplying a single product or service and not an overall experience. So, each part is suboptimized. These systems are also designed by humans, but it is certainly not easy to change these gradually grown systems. Apart from the context and the system, a product or system should also fit to the human capabilities. Applying human factors and ergonomics in the design can certainly be of help in the best fit to human capabilities. The International Ergonomics Association (www.iea.cc) distinguishes three domains:

- Physical ergonomics, which is concerned with human anatomical, anthropometric, physiological and biomechanical characteristics.
- Cognitive ergonomics, which is concerned with mental processes, such as perception, memory and reasoning, and
- Organizational ergonomics, which is concerned with the optimization of sociotechnical systems.

7.1 PHYSICAL ERGONOMICS

In this book, the importance of using physical ergonomics in designing is shown in various chapters. Anthropometrics, the discipline that studies the sizes of the different parts of the human body, is relevant in many chapters. In designing the sleep train, the position humans assume while we sleep is relevant, and by combining the posture with the sizes of the different body parts, we can calculate the bed size that is needed to accommodate these postures. But it is also relevant in replacing the duvet cover. We can calculate the maximum width between our hands standing with the arms spread and easily estimate which sizes can be handled by humans. And in designing the seats, we can check what the hip breadth is of different populations (e.g., in www.dined.nl) and what the elbow-elbow width is. Using anthropometric tables is not always enough. Usually, we need to observe the behavior. For instance,

DOI: 10.1201/9781003637035-7

Epilogue

The title of this book is *50 Ways to Fool the User*, and we do have 50 chapters, but actually we have only described 49 products – fooled you!

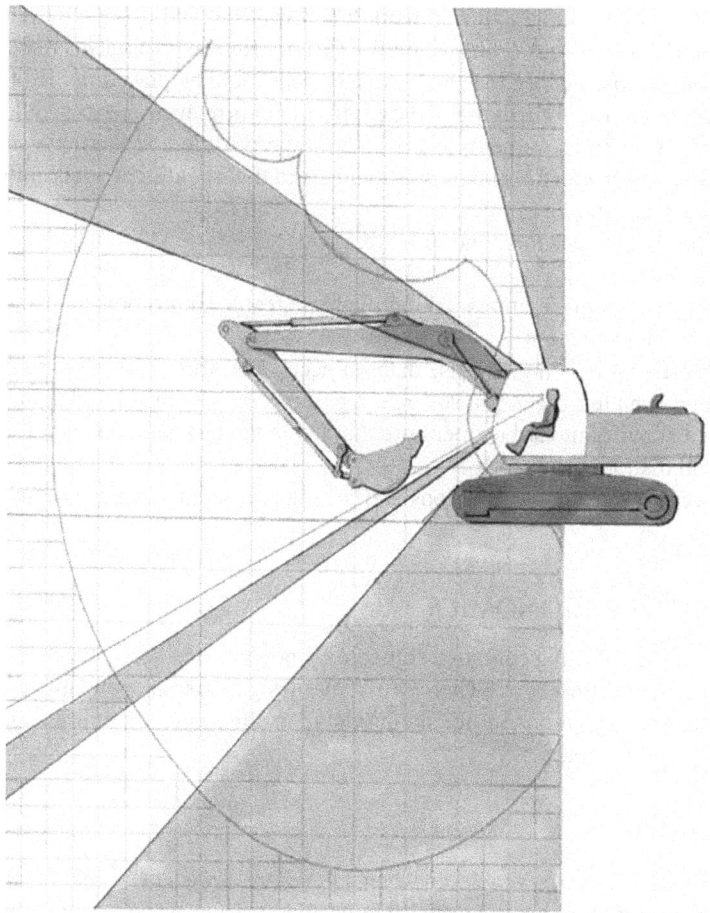

FIGURE 7.1

in designing a flight deck of an airplane, it is not enough to check the arm length if pilots can reach the buttons. We also need to check how the pilots grab and touch the buttons as it influences the position of the joints and thereby the reach envelope.

A part of physical ergonomics is also human physical capabilities. Ergonomic research is needed when acceptable lifting and pushing-pulling forces are to be determined for working conditions, but also for opening a packaging the human capabilities should be taken into account.

Environmental characteristics like temperature and noise are also related to physical ergonomics. Noise should not be that loud or that type of sound that it distracts from work. Adapting the environmental temperature to the tasks has an influence on the performance of employees and their comfort, which makes attention to temperature very relevant. An aspect gaining more attention is stimulating movement. In many chapters, preventing long static situations, where humans do not move it is stated that it should be avoided. During work lunch walking and using a bicycle at the desk is promoted. This has influence on the design of the environment. It means that around offices a park is needed and bicycle desks should be placed in the office building. Also, the choice of furniture in the environment is relevant. Seats should be chosen that enable small movements to avoid the static posture and facilitate small movements. Also, environmental interiors should avoid that standing work is obliged the whole day as it should be avoided.

The tips regarding physical ergonomics are:

- design products and situations, taking the human size (anthropometry) and reachability into account.
- when human force is needed, use data on acceptable lifting and pull-push forces, and don't forget acceptable forces regarding the opening of packaging.
- arrange an environmental temperature that fits to the task and reduce noise that could distract form work
- try to prevent long static situations and arrange variations in posture and activities.

7.2   COGNITIVE ERGONOMICS

Also, the importance of using cognitive ergonomic knowledge in designing is shown in various chapters of this book. There are some design principles mentioned by Norman (2006) that are very true for the products in this book: visibility, consistency, mapping and feedback.

7.2.1   VISIBILITY

When the toilet paper is not visible or when the controls are not visible or when the button for starting the car is located behind the steering wheel, humans have difficulty in using the device. It is amazing how many products the essential elements for controlling or understanding the product or system are out of sight. This is certainly something that can be improved in many products. In designing, it can be tested based on paper. In the figure in this chapter, a side view of the visible areas of an earth-moving machine is shown. This is a way to study visibility, but of course enough light on the part of the object that should be touched or handled by the human is also essential. The handles under the seat pan of an office chair are difficult to see from the seated position and often in the dark, which makes using them more difficult.

7.2.2   CONSISTENCY

Using the sink drain, starting a car or using the menu of a call center, it is usually a different experience than in previous systems you used. This is inconsistency. Making a product consistent with its previous products or products we are used to makes it much easier for the user.

7.2.3 FEEDBACK

Immediate feedback is essential. This was described in, for instance, the remote control chapter. If you push a button and nothing happens, users will think they did not set something in action. So it is very helpful if activational feedback is provided because it shows the user that something has happened.

7.2.4 MEMORY

The chapter on passwords shows that the human capacity to remember things is limited. This should be taken into account in, for instance, the number of numbers and special signs in a password.

7.3 ORGANIZATIONAL ERGONOMICS

In the chapter on management and the chapter on control, we describe that humans like to be in control. How you organize the work and activities of humans is the area of organizational ergonomics. Also, how you design an office interior and separate departments, which make noise (advertisement) from departments that need to work concentrated (e.g. journalists), is part of the organizational ergonomics.

7.4 THE FUTURE

So how can we improve the ergonomic design of products and systems? Here are some key principles to consider for good ergonomic design.

1. **User-centered**: the design should satisfy the needs, preferences, ease of use and any physical and cognitive limitations of the end users. This includes accessibility considerations for those with different abilities and disabilities.
2. **Accessibility**: the designs should be usable by people of all physical and mental abilities, including those with disabilities.
3. **Safety and comfort**: the design should be safe and comfortable to use. It should minimize physical strain and maximize user comfort by considering factors like posture, reach, and movement. It should be safe to use and encourage users to maintain natural, neutral postures to reduce muscle strain and fatigue.
4. **Simplicity**: the design should be simple and intuitive to use. It should reduce the cognitive load on users by using clear, intuitive displays and controls, and avoid unnecessary actions.
5. **Consistency**: the design elements should maintain consistency across products to help users learn and use the system more easily. Maintain consistent design patterns and terminology throughout the interface can reduce cognitive load. This is why we have, for instance, internationally agreed symbols for products.
6. **Displays and controls**: the information displays and controls should optimize good signal detection and appropriate control response. If the display

is visual, then text, icons, and other visual elements should be easily readable and distinguishable, and appropriate font sizes, contrasts and should be used. If it is auditory, then the sound should be louder than background noise and be an alerting frequency. If it is a warning of imminent danger, then it should be intermittent, such as a flashing light and a wailing siren.

7. **Effort**: the design should maximize efficiency and minimize effort (unless it's a gym where you are lifting weights). Users should be able to use the design quickly and accurately with minimal physical and mental effort.

8. **Feedback**: the design should provide clear and immediate feedback to users about their actions to help them understand the results of their interactions.

9. **Flexibility and customization**: where possible, the design should allow for customization and adaptability to accommodate different user preferences and needs.

10. **Accuracy**: designs should minimize the potential for errors and provide clear recovery options.

11. **Fun**: where appropriate, designs should be enjoyable to use. If it's fun, you will use it again!

12. **Environmental factors**: designs should account for factors like lighting, noise levels, and temperature, as they can impact user comfort and productivity when using the design.

13. **Variation of posture**: designs should support humans to vary their posture and systems should support human to avoid long standing and sitting.

For all fields, it is important to mention that there is not enough knowledge to predict for every system and product, how it can be user-friendly. Therefore, our advice is to have a user test in the design phase or prototyping and in the final product. Also, a cognitive walkthrough and simulation with models could be applied to make a better product. Choosing the right test population is important. It should reflect the real end-user. For computer software that will be used by farmers, office clerks and people working in households, a test among computer experts is certainly not the correct way to do it.

REFERENCE

Dougali, E., Santema, S. C., & Beelaerts van Blokland, W. W. A. (2014). *How to design a service for air-traveling from a co-creating perspective: A case study for Athens, Greece & Amsterdam, Netherlands*. Proceedings of TMCE 2014, pp. 1–13.

Norman, D. (2006). *The design of everyday things: Revised and expanded edition*. Basic books.